BATTLES WITH THE SEA

LECTOR HOUSE PUBLIC DOMAIN WORKS

BATTLES WITH THE SEA

ROBERT MICHAEL BALLANTYNE

ISBN: 978-93-90294-73-2

Published: -

LECTOR HOUSE LLP

LECTOR HOUSE LLP
E-MAIL: lectorpublishing@gmail.com

"BATTLES WITH THE SEA"
HEROES OF THE LIFEBOAT AND ROCKET

BY

R. M. BALLANTYNE

AUTHOR OF "DUSTY DIAMONDS;" "THE BATTERY AND THE BOILER;"
"THE GIANT OF THE NORTH;" "THE LONELY ISLAND;" "POST HASTE;
A TALE OF HER MAJESTY'S MAILS;" "IN THE TRACK OF THE TROOPS;"
"THE SETTLER AND THE SAVAGE;" "UNDER THE WAVES;"
"RIVERS OF ICE;" "BLACK IVORY;" "THE PIRATE CITY;" "THE
NORSEMEN IN THE WEST;" "THE IRON HORSE;"
"THE FLOATING LIGHT OF THE GOODWIN SANDS;"
"ERLING THE BOLD;" "FIGHTING THE FLAMES;"
"SHIFTING WINDS;" "DEEP DOWN;" "THE LIGHTHOUSE;"
"GASCOYNE ;" "THE LIFE BOAT;"
"THE GOLDEN DREAM," ETC.

WITH ILLUSTRATIONS

PREFACE.

The only war which deserves the name of "glorious" is fought—has been and will be fought to the end of time—on our sea-coast, and some of our most notable victories have been gained there.

The men and material with which this war is waged form the subject of this little book, together with a few observations more or less weighty.

R. M. B.

Harrow-on-the-Hill,
1883.

CONTENTS

BATTLES WITH THE SEA

LEAVING THE GOODWIN SANDS—VICTORIOUS.

CHAPTER ONE.

SKIRMISHES WITH THE SUBJECT GENERALLY.

It ought to be known to all English boys that there is a terrible and costly war in which the British nation is at all times engaged. No intervals of peace mark the course of this war. Cessations of hostilities there are for brief periods, but no treaties of peace. "War to the knife" is its character. Quarter is neither given nor sought. Our foe is unfeeling, unrelenting. He wastes no time in diplomatic preliminaries; he scorns the courtesies of national life. No ambassadors are recalled, no declarations of war made. Like the Red Savage he steals upon us unawares, and, with a roar of wrathful fury, settles down to his deadly work.

How does this war progress? It is needful to put and reiterate this question from time to time, because new generations of boys are always growing up, who, so far from being familiar with the stirring episodes of this war, and the daring deeds of valour performed, scarcely realise the fact that such a war is being carried on at all, much less that it costs hundreds of lives and millions of money every year.

It may be styled a naval war, being waged chiefly in boats upon the sea. It is a war which will never cease, because our foe is invincible, and we will never give in; a war which, unlike much ordinary warfare, is never unjust or unnecessary; which cannot be avoided, which is conducted on the most barbarous principles of deathless enmity, but which, nevertheless, brings true glory and honour to those heroes who are ever ready, night and day, to take their lives in their hands and rush into the thick of the furious fray.

Although this great war began—at least in a systematic manner—only little more than fifty years ago, it will not end until the hearts of brave and generous Britons cease to beat, and the wild winds cease to blow, for the undying and unconquerable enemy of whom we write is—the Storm!

"Death or victory!" the old familiar warwhoop, is not the final war-cry here. Death is, indeed, always faced—sometimes met—and victory is often gained; but, final conquests being impossible, and the "piping times of peace" being out of the question, the signal for the onset has been altered, and the world's old battle-cry has been exchanged for the soul-stirring shout of "Rescue the perishing!"

DRIVING BEFORE THE STORM.

Though our foe cannot be slain, he can, like the genii of Eastern story, be baffled.

In the days of old, the Storm had it nearly all his own way. Hearts, indeed, were not less brave, but munitions of war were wanting. In this matter, as in everything else, the world is better off now than it was then. Our weapons are more perfect, our engines more formidable. We can now dash at our enemy in the very heart of his own terrible strongholds; fight him where even the boldest of the ancient Vikings did not dare to venture, and rescue the prey from the very jaws of death amid the scenes of its wildest revelry.

The heroes who recruit the battalions of our invincible army are the bronzed and stalwart men of our sea-coast towns, villages, and hamlets—men who have had much and long experience of the foe with whom they have to deal. Their panoply is familiar to most of us. The helmet, a sou'wester; the breastplate, a lifebelt of cork; the sword, a strong short oar; their war-galley, a splendid *lifeboat*; and their shield—the Hand of God.

In this and succeeding chapters I purpose to exhibit and explain in detail our Lifeboats, and the great, the glorious work which they annually accomplish; also the operations of the life-saving Rocket, which has for many years rescued innumerable lives, where, from the nature of circumstances, Lifeboats could not have gone into action. I hold that we—especially those of us who dwell in the interior of our land—are not sufficiently alive to the deeds of daring, the thrilling incidents, the terrible tragedies and the magnificent rescues which are perpetually going on around our shores. We are not sufficiently impressed, perhaps, with the *nationality* of the work done by the Royal National Lifeboat Institution, which manages our fleet of 270 lifeboats. We do not fully appreciate, it may be, the personal interest which we ourselves have in the great war, and the duty—to say nothing of privilege—which lies upon us to lend a helping hand in the good cause.

Before going into the marrow of the subject, let us put on the wings of imagination, and soar to such a height that we shall be able to take in at one eagle glance all the coasts of the United Kingdom—a sweep of about 5000 miles all round! It is a tremendous sight, for a storm is raging! Black clouds are driving across the murky sky; peals of thunder rend the heavens; lightning gleams at intervals, revealing more clearly the crested billows that here roar over the sands, or there churn and seethe among the rocks. The shrieking gale sweeps clouds of spray high over our windward cliffs, and carries flecks of foam far inland, to tell of the dread warfare that is raging on the maddened sea.

Near the shore itself numerous black specks are seen everywhere, like ink-spots on the foam. These are wrecks, and the shrieks and the despairing cries of the perishing rise above even the roaring of the gale. Death is busy, gathering a rich harvest, for this is a notable night in the great war. The Storm-fiend is roused. The enemy is abroad in force, and has made one of his most violent assaults, so that from Shetland to Cornwall, ships and boats are being battered to pieces on the rocks and sands, and many lives are being swallowed up or dashed out; while, if you turn your gaze further out to sea, you will descry other ships and boats and victims hurrying onward to their doom. Here, a stately barque, with disordered topsails almost bursting from the yards as she hurries her hapless crew—all ignorant, perchance, of its proximity—towards the dread lee-shore. Elsewhere, looming through the murk, a ponderous merchantman, her mainmast and mizzen gone, and just enough of the foremast left to support the bellying foresail that bears her to destruction.

Think you, reader, that this sketch is exaggerated? If so, let us descend from our lofty outlook, and take a nearer view of facts in detail. I quote the substance of the following from a newspaper article published some years ago.

AT THE MERCY OF WIND AND WAVE.

The violence of the storm on Wednesday and Thursday night was terrific. The damage to shipping has been fearful. On sea the tremendous gale proved disastrous beyond precedent. Falmouth Harbour was the scene of several collisions, and one barque and a tug steamer sank at their anchors. A wreck is reported at Lelant, to which the Penzance lifeboat with a stout-hearted crew had started, when our despatch left, to rescue thirteen men who could be descried hanging in the shrouds. A fine new ship is on Hayle bar, and another vessel is believed to be wrecked there also. Doubtless we have not yet heard of all the wrecks on the Cornish coast; but it is in the magnificent bay which includes Torquay, Paignton, and Brixham that the most terrible havoc has occurred. On Wednesday, about sixty sail were anchored in Torbay. Eleven have gone ashore at Broadsands, five of which are total wrecks. The names of those we could ascertain were the Fortitude, of Exeter; the Stately, of Newcastle; the Dorset, of Falmouth, and a French brigantine. At five o'clock on Thursday evening some of the crews were being drawn ashore by lines and baskets. At Churston Cove one schooner is ashore and a total wreck; there is also another, the Blue Jacket, which may yet be saved. At Brixham there are two fine ships ashore inside the breakwater. At the back of the pier ten vessels have been pounded to matchwood, and all that remains are a shattered barque, her masts still standing, two brigs, and a schooner, all inextricably mingled together. Twelve trawlers have been sunk and destroyed. Out of the sixty ships at anchor on Wednesday night there were not more than ten left on Thursday afternoon. Many of these are disabled, some dismasted. A fishing-boat belonging to Brixham was upset in the outer harbour about eight o'clock, and two married fishermen of the town and a boy were drowned. At Elbury a new brig, the Zouave, of Plymouth, has gone to pieces, and six out of her crew of ten are drowned. Eleven other vessels are on shore at Elbury, many of the men belonging to which cannot be accounted for. One noble woman, named Wheaton, wife of a master mariner, saved two lives by throwing a rope from the window of her house, which is built on the rocks overhanging the bay at Furzeham Hill. Scores of poor shipwrecked men are wandering distractedly about Brixham and Churston, the greater part of them having lost all they possessed. The total loss of life arising from these disasters is variously estimated at from seventy to a hundred.

Is not this a tremendous account of the doings of one gale? And let it be observed that we have lifted only one corner of the curtain and revealed the battle-field of only one small portion of our far-reaching coasts. What is to be said of the other parts of our shores during that same wild storm? It would take volumes instead of chapters to give the thrilling incidents of disaster and heroism in full detail. To convey the truth in all its force is impossible, but a glimmering of it may be obtained by a glance at the Wreck Chart which is published by the Board of Trade every year.

Every black spot on that chart represents a wreck more or less disastrous, which occurred in the twelve months. It is an appalling fact that about two thousand ships, upwards of seven hundred lives, and nearly two millions sterling, are lost *every* year on the shores of the United Kingdom. Some years the loss is heavier, sometimes lighter, but in round numbers this is our annual loss in the great war. That it would be far greater if we had no lifeboats and no life-saving rockets it will be our duty by-and-by to show.

The black spots on the Wreck Chart to which we have referred show at a single glance that the distribution of wrecks is very unequal—naturally so. Near the great seaports we find them thickly strewn; at other places, where vessels pass in great numbers on their way to these ports, the spots are also very numerous, while on unfrequented parts they are found only here and there in little groups of two, three, or four. Away on the nor'-west shores of Scotland, for instance, where the seal and the sea-mew have the ocean and rugged cliffs pretty much to themselves, the plague-spots are few and far between; but on the east coast we find a fair sprinkling of them, especially in the mouths of the Forth and Tay, whither a goodly portion of the world's shipping crowds, and to which the hardy Norseman now sends many a load of timber—both log and batten—instead of coming, as he did of old, to batten on the land. It is much the same with Ireland, its more important seaports being on the east.

TO THE RESCUE!

But there is a great and sudden increase of the spots when we come to England. They commence at the border, on the west, where vessels from and to the busy Clyde enter or quit the Irish Sea. Darkening the fringes of the land on both sides, and clustering round the Isle of Man, they multiply until the ports have no room to hold them, and, as at Liverpool, they are crowded out into the sea. From the deadly shores of Anglesea, where the Royal Charter went down in the great and memorable storm of November, 1859, the signs of wreck and disaster thicken as we go south until we reach the Bristol Channel, which appears to be choked with them, and the dangerous cliffs of Cornwall, which receive the ill-fated vessels of the fleets that are perpetually leaving or entering the two great channels. But it is on the east coast of England that the greatest damage is done. From Berwick to the Thames the black spots cluster like bees. On the coasts of Norfolk and Suffolk, off Great Yarmouth, where lie the dangerous Haisborough Sands, the spots are no longer in scattered groups, but range themselves in dense battalions; and further south, off the coast of Kent, round which the world's commerce flows unceasingly into the giant metropolis, where the famous Goodwin Sands play their deadly part in the great war, the dismal spots are seen to cluster densely, like gnats in a summer sky.

Now, just where the black spots are thickest on this wreck chart, lifeboats and rocket apparatus have been stationed in greatest numbers. As in ordinary warfare, so in battles with the sea, our "Storm Warriors"[1] are thrown forward in force where the enemy's assaults are most frequent and dangerous. Hence we find the eastern shores of England crowded at every point with life-saving apparatus, while most of the other dangerous parts of the coast are pretty well guarded.

Where and how do our coast heroes fight? I answer—sometimes on the cliffs, sometimes on the sands, sometimes on the sea, and sometimes even on the pier-heads. Their operations are varied by circumstances. Let us draw nearer and look at them while in action, and observe how the enemy assails them. I shall confine myself at present to a skirmish.

When the storm-fiend is abroad; when dark clouds lower; when blinding rain or sleet drives before the angry gale, and muttering thunder comes rolling over the sea, men with hard hands and weather-beaten faces, clad in oilskin coats and sou'-westers, saunter down to our quays and headlands all round the kingdom. These are the lifeboat crews and rocket brigades. They are on the lookout. The enemy is moving, and the sentinels are being posted for the night—or rather, they are posting themselves, for nearly all the fighting men in this war are volunteers!

They require no drilling to prepare them for the field; no bugle or drum to sound the charge. Their drum is the rattling thunder; their trumpet the roaring storm. They began to train for this warfare when they were not so tall as their fathers' boots, and there are no awkward squads among them now. Their organisation is rough-and-ready, like themselves, and simple too. The heavens call them to action; the coxswain grasps the helm, the oars are manned, the word is given, and the rest is straightforward fighting—over everything, through everything, in the teeth of everything, until the victory is gained, and rescued men, women, and children are landed in safety on the shore.

Of course they do not always succeed, but they seldom or never fail to do the very uttermost that it is in the power of strong and daring men to accomplish. Frequently they can tell of defeat and victory on the same battlefield.

So it was on one fearful winter night at the mouth of the Tyne in the year 1867. The gale that night was furious. It suddenly chopped round to the South South East, and, as if the change had recruited its energies, it blew a perfect hurricane between midnight and two in the morning, accompanied by blinding showers of sleet and hail, which seemed to cut like a knife. The sea was rising mountains high.

About midnight, when the storm was gathering force and the sentinels were scarcely able to keep a lookout, a preventive officer saw a vessel driving ashore to the south of the South Pier. Instantly he burnt a blue light, at which signal three guns were fired from the Spanish Battery to call out the Life Brigade. The men

[1] See an admirable book, with this title, written by the Reverend John Gilmore, of Ramsgate. (Macmillan and Company)

were on the alert. About twenty members of the brigade assembled almost immediately on the pier, where they found that the preventive officer and pier-police-
man had already got out the life-saving apparatus; but the gale was so fierce that
they had been forced to crawl on their hands and knees to do so. A few minutes
more and the number of brigade men increased to between fifty and sixty. Soon
they saw, through the hurtling storm, that several vessels were driving on shore.
Before long, four ships, with their sails blown to ribbons, were grinding themselves to powder, and crashing against each other and the pier-sides in a most
fearful manner. They were the Mary Mac, the Cora, and the Maghee, belonging to
Whitstable, and the Lucern of Blyth.

Several lifeboats were stationed at that point. They were all launched, manned,
and promptly pulled into the Narrows, but the force of the hurricane and seas
were such that they could not make headway against them. The powers of man
are limited. When there is a will there is not always a way! For two hours did these
brave men strain at the tough oars in vain; then they unwillingly put about and
returned, utterly exhausted, leaving it to the men with the life-lines on shore to
do the fighting. Thus, frequently, when one arm of the service is prevented from
acting; the other arm comes into play.

The work of the men engaged on the pier was perilous and difficult, for the
lines had to be fired against a head wind. The piers were covered with ice, and the
gale was so strong that the men could hardly stand, while the crews of the wrecks
were so benumbed that they could make little effort to help themselves.

The men of the Mary Mac, however, made a vigorous effort to get their long-
boat out. A boy jumped in to steady it. Before the men could follow, the boat was
stove in, the rope that held it broke, and it drove away with the poor lad in it. He
was quickly washed out, but held on to the gunwale until it drifted into broken
water, when he was swallowed by the raging sea and the boat was dashed to pieces.

Meanwhile the crew of the Cora managed to swing themselves ashore, their
vessel being close to the pier. The crew of the Lucern, acting on the advice of
the brigade men, succeeded in scrambling on board the Cora and were hauled
ashore on the life-lines. They had not been ten minutes out of their vessel when
she turned over with her decks towards the terrible sea, which literally tore her
asunder, and pitched her up, stem on end, as if she had been a toy. The crew of the
Maghee were in like manner hauled on to the pier, with the exception of one lad
from Canterbury. It was the poor boy's first voyage. Little did he think probably,
while dreaming of the adventures of a sailor's career, what a terrible fate awaited
him. He was apparently paralysed with fear, and could not spring after his com-
rades to the pier, but took to the rigging. He had scarcely done so when the vessel
heeled over, and he was swung two or three times backwards and forwards with
the motion of the masts.

It is impossible to imagine the feelings of the brave men on the pier, who
would so gladly have risked their lives to save him—he was so near, and yet so
hopelessly beyond the reach of human aid!

In a very brief space of time the waves did their work—ship and boy were swallowed up together.

While these events were enacting on the pier the Mary Mac had drifted over the sand about half a mile from where she had struck. One of her crew threw a leadline towards a seaman on the shore. The hero plunged into the surf and caught it. The rest of the work was easy. By means of the line the men of the Life Brigade sent off their hawser, and breeches-buoy or cradle (which apparatus I shall hereafter explain), and drew the crew in safety to the land.

That same morning a Whitby brig struck on the sands. The lifeboat Pomfret, belonging to the Royal National Lifeboat Institution, put out and rescued her crew. In the morning the shores were strewn with wreckage, and amongst it was found the body of the boy belonging to the Mary Mac.

All these disasters were caused by the masters of the vessels mistaking the south for the north pier, in consequence of having lost sight of Tynemouth light in the blinding showers.

Of course many lifeboats were out doing good service on the night to which I have referred, but I pass all that by at present. The next chapter will carry you, good reader, into the midst of a pitched battle.

CHAPTER TWO.

DESCRIBES A TREMENDOUS BATTLE AND A GLORIOUS VICTORY.

Before following our brilliant lifeboat—this gaudy, butterfly-like thing of red, white, and blue—to the field of battle, let me observe that the boats of the Royal National Lifeboat Institution have several characteristic qualities, to which reference shall be made hereafter, and that they are of various sizes.[2]

One of the largest size is that of Ramsgate. This may be styled a privileged boat, for it has a steam-tug to wait upon it named the Aid. Day and night the Aid has her fires "banked up" to keep her boilers simmering, so that when the emergency arises, a vigorous thrust of her giant poker brings them quickly to the boiling point, and she is ready to take her lifeboat in tow and tug her out to the famed and fatal Goodwin Sands, which lie about four miles off the coast—opposite to Ramsgate.

I draw attention to this boat, first because she is exceptionally situated with regard to frequency of call, the means of going promptly into action, and success in her work. Her sister-lifeboats of Broadstairs and Margate may, indeed, be as often called to act, but they lack the attendant steamer, and sometimes, despite the skill and courage of their crews, find it impossible to get out in the teeth of a tempest with only sail and oar to aid them.

Early in December, 1863, an emigrant ship set sail for the Antipodes; she was the Fusilier, of London. It was her last voyage, and fated to be very short. The shores of Old England were still in sight, the eyes of those who sought to "better their circumstances" in Australia were yet wet, and their hearts still full with the grief of parting from loved ones at home, when one of the most furious storms of the season caught them and cast their gallant ship upon the dangerous Sands off the mouth of the Thames. This happened on the night of the 3rd, which was intensely dark, as well as bitterly cold.

Who can describe or conceive the scene that ensued! the horror, the shrieking of women and children, and the yelling of the blast through the rigging,—for it was an absolute hurricane,—while tons of water fell over the decks continually, sweeping them from stem to stern.

[2] A full and graphic account of the Royal National Lifeboat Institution—its boats, its work, and its achievements—may be found in an interesting volume by its late secretary, Richard Lewis, Esquire, entitled *History of the Lifeboat and its Work*—published by Macmillan and Company.

The Fusilier had struck on that part of the sands named the Girdler. In the midst of the turmoil there was but one course open to the crew—namely, to send forth signals of distress. Guns were fired, rockets sent up, and tar-barrels set a-blaze. Then, during many hours of agony, they had to wait and pray.

On that same night another good ship struck upon the same sands at a different point—the Demerara of Greenock—not an emigrant ship, but freighted with a crew of nineteen souls, including a Trinity pilot. Tossed like a plaything on the Sands—at that part named the Shingles—off Margate, the Demerara soon began to break up, and the helpless crew did as those of the Fusilier had done and were still doing—they signalled for aid. But it seemed a forlorn resource. Through the thick, driving, murky atmosphere nothing but utter blackness could be seen, though the blazing of their own tar-barrels revealed, with awful power, the seething breakers around, which, as if maddened by the obstruction of the sands, leaped and hissed wildly over them, and finally crushed their vessel over on its beam-ends. Swept from the deck, which was no longer a platform, but, as it were, a sloping wall, the crew took refuge in the rigging of one of the masts which still held fast. The mast overhung the caldron of foam, which seemed to boil and leap at the crew as if in disappointed fury.

By degrees the hull of the Demerara began to break up. Her timbers writhed and snapped under the force of the ever-thundering waves as if tormented. The deck was blown out by the confined and compressed air. The copper began to peel off, the planks to loosen, and soon it became evident that the mast to which the crew were lashed could not long hold up. Thus, for ten apparently endless hours the perishing seamen hung suspended over what seemed to be their grave. They hung thus in the midst of pitchy darkness after their blazing tar-barrels had been extinguished.

And what of the lifeboat-men during all this time? Were they asleep? Nay, verily! Everywhere they stood at pierheads, almost torn from their holdfasts by the furious gale, or they cowered under the lee of boats and boat-houses on the beach, trying to gaze seaward through the blinding storm, but nothing whatever could they see of the disasters on these outlying sands.

There are, however, several sentinels which mount guard night and day close to the Goodwin and other Sands. These are the Floating Lights which mark the position of our extensive and dangerous shoals. Two of these sentinels, the Tongue lightship and the Prince's lightship, in the vicinity of the Girdler Sands, saw the signals of distress. Instantly their guns and rockets gleamed and thundered intelligence to the shore. Such signals had been watched for keenly that night by the brave men of the Margate lifeboat, who instantly went off to the rescue. But there are conditions against which human courage and power and will are equally unavailing. In the teeth of such a gale from the west-nor'-west, with the sea driving in thunder straight on the beach, it was impossible for the Margate boat to put out. A telegram was therefore despatched to Ramsgate. Here, too, as at Broadstairs, and everywhere else, the heroes of the coast were on the lookout, knowing well the duties that might be required of them at any moment.

The stout little Aid was lying at the pier with her steam "up." The Ramsgate lifeboat was floating quietly in the harbour, and her sturdy lion-like coxswain, Isaac Jarman, was at the pier-head with some of his men, watching. The Ramsgate men had already been out on service at the sands that day, and their appetite for saving life had been whetted. They were ready for more work. At a quarter past eight p.m. the telegram was received by the harbour-master. The signal was given. The lifeboat-men rushed to their boats.

"First come, first served," is the rule there. She was over-manned, and some of the brave fellows had to leave her. The tight little tug took the boat in tow, and in less than half an hour rushed out with her into the intense darkness, right in the teeth of tempest and billows.

The engines of the Aid are powerful, like her whole frame. Though fiercely opposed she battled out into the raging sea, now tossed on the tops of the mighty waves, now swallowed in the troughs between. Battered by the breaking crests, whelmed at times by "green seas," staggering like a drunken thing, and buffeted by the fierce gale, but never giving way an inch, onward, steadily if slowly, until she rounded the North Foreland. Then the rescuers saw the signals going up steadily, regularly, from the two lightships. No cessation of these signals until they should be answered by signals from the shore.

All this time the lifeboat had been rushing, surging, and bounding in the wake of her steamer. The seas not only roared around her, but absolutely overwhelmed her. She was dragged violently over them, and sometimes right through them. Her crew crouched almost flat on the thwarts, and held on to prevent being washed overboard. The stout cable had to be let out to its full extent to prevent snapping, so that the mist and rain sometimes prevented her crew from seeing the steamer, while cross seas met and hurled her from side to side, causing her to plunge and kick like a wild horse.

About midnight the Tongue lightship was reached and hailed. The answer given was brief and to the point: "A vessel in distress to the nor'-west, supposed to be on the high part of the Shingles Sand!"

Away went the tug and boat to the nor'-west, but no vessel could be found, though anxious hearts and sharp and practised eyes were strained to the uttermost. The captain of the Aid, who knew every foot of the sands, and who had medals and letters from kings and emperors in acknowledgment of his valuable services, was not to be balked easily. He crept along as close to the dangerous sands as was consistent with the safety of his vessel.

How intently they gazed and listened both from lifeboat and steamer, but no cry was to be heard, no signal of distress, nothing but the roaring of the waves and shrieking of the blast, and yet they were not far from the perishing! The crew of the Demerara were clinging to their quivering mast close by, but what could their weak voices avail in such a storm? Their signal fires had long before been drowned out, and those who would have saved them could not see more than a few yards around.

Presently the booming of distant cannon was heard and then a faint line of fire was seen in the far distance against the black sky. The Prince's and the Girdler lightships were both firing guns and rockets to tell that shipwreck was taking place near to them. What was to be done? Were the Shingles to be forsaken, when possibly human beings were perishing there? There was no help for it. The steamer and lifeboat made for the vessels that were signalling, and as the exhausted crew on the quivering mast of the Demerara saw their lights depart, the last hope died out of their breasts.

"Hope thou in God, for thou shalt yet praise Him," perchance occurred to some of them: who knows?

Meanwhile the rescuers made for the Prince's lightship and were told that a vessel in distress was signalling on the higher part of the Girdler Sands.

Away they went again, and this time were successful. They made for the Girdler lightship, and on the Girdler Sands they found the Fusilier.

The steamer towed the lifeboat to windward of the wreck into such a position that when cast adrift she could bear down on her. Then the cable was slipped and the boat went in for her own special and hazardous work. Up went her little foresail close-reefed, and she rushed into a sea of tumultuous broken water that would have swamped any other kind of boat in the world.

What a burst of thrilling joy and hope there was among the emigrants in the Fusilier when the little craft was at last descried! It was about one o'clock in the morning by that time, and the sky had cleared a very little, so that a faint gleam of moonlight enabled them to see the boat of mercy plunging towards them through a very chaos of surging seas and whirling foam. To the rescuers the wreck was rendered clearly visible by the lurid light of her burning tar-barrels as she lay on the sands, writhing and trembling like a living thing in agony. The waves burst over her continually, and, mingling in spray with the black smoke of her fires, swept furiously away to leeward.

At first each wave had lifted the ship and let her crash down on the sands, but as the tide fell this action decreased, and had ceased entirely when the lifeboat arrived.

And now the point of greatest danger was reached. How to bring a lifeboat alongside of a wreck so as to get the people into her without being dashed to pieces is a difficult problem to solve. It was no new problem, however, to these hardy and fearless men; they had solved it many a time, before that night. When more than a hundred yards to windward of the wreck, the boat's foresail was lowered and her anchor let go. Then they seized the oars, and the cable was payed out; but the distance had been miscalculated. They were twenty yards or so short of the wreck when the cable had run completely out, so the men had to pull slowly and laboriously back to their anchor again, while the emigrants sent up a cry of despair, supposing they had failed and were going to forsake them! At length the anchor was got up. In a few minutes it was let go in a better position, and the boat was carefully veered down under the lee of the vessel, from both bow and stern of which a hawser was thrown to it and made fast. By means of these ropes and the cable the boat was kept somewhat in position without striking the wreck.

It was no easy matter to make the voice heard in such a gale and turmoil of seas, but the captain of the Fusilier managed to give his ship's name and intended destination. Then he shouted, "How many can you carry? We have more than a hundred souls on board; more than sixty of them women and children."

This might well fill the breasts of the rescuers with anxiety. Their boat, when packed full, could only carry about thirty. However, a cheering reply was returned, and, seizing a favourable opportunity, two of the boatmen sprang on the wreck, clambered over the side, and leaped among the excited emigrants. Some seized them by the hands and hailed them as deliverers; others, half dead with terror, clung to them as if afraid they might forsake them. There was no time, however, to humour feelings. Shaking them all off—kindly but forcibly—the men went to work with a will, briefly explained that there was a steamer not far off, and began to get the women first into the boat.

Terror-stricken, half fainting, trembling in every limb, deadly pale, and exhausted by prolonged anxiety and exposure, the poor creatures were carried rather than led to the ship's side. It needed courage even to submit to be saved on such a night and in such circumstances. Two sailors stood outside the ship's bulwarks, fastened there by ropes, ready to lower the women. At one moment the raging sea rose with a roar almost to the feet of these men, bearing the kicking lifeboat on its crest. Next moment the billow had passed, and the men looked down into a yawning abyss of foam, with the boat surging away far out of their reach, plunging and tugging at the ropes which held it, as a wild horse of the plains might struggle with the lasso. No wonder that the women gazed appalled at the prospect of such a leap, or that some shrieked and wildly resisted the kind violence of their rescuers. But the leap was for life; it had to be taken—and quickly, too, for the storm was very fierce, and there were many to save!

One of the women is held firmly by the two men. With wildly-staring eyes she sees the boat sweep towards her on the breast of a rushing sea. It comes closer. Some of the men below stand up with outstretched arms. The woman makes a half spring, but hesitates. The momentary action proves almost fatal. In an instant the boat sinks into a gulf, sweeps away as far as the ropes will let her, and is buried in foam, while the woman is slipping from the grasp of the men who hold her.

"Don't let her go! don't let her go!" is roared by the lifeboat-men, but she has struggled out of their grasp. Another instant and she is gone; but God in His mercy sends the boat in again at that instant; the men catch her as she falls, and drag her inboard.

Thus, one by one, were the women got into the lifeboat. Some of these women were old and infirm; some were invalids. Who can conceive the horror of the situation to such as these, save those who went through it?

The children were wrapped up in blankets and thus handed down. Some of the husbands or fathers on board rolled up shawls and blankets and tossed them down to the partially clothed and trembling women. It chanced that one small infant was bundled up in a blanket by a frantic passenger and handed over the side. The man who received it, mistaking it for merely a blanket, cried, "Here, Bill, catch!" and tossed it into the boat. Bill, with difficulty, caught it as it was flying overboard; at the same moment a woman cried, "My child! my child!" sprang forward, snatched the bundle from the horrified Bill, and hugged it to her bosom!

At last the boat, being sufficiently filled, was hauled up to her anchor. Sail was hoisted, and away they flew into the surging darkness, leaving the rest of the emigrants still filled with terrible anxiety, but not now with hopeless despair.

The lifeboat and her tender work admirably together. Knowing exactly what must be going on, and what would be required of him, though he could see nothing, the captain of the Aid, after the boat had slipped from him, had run down along the sands to leeward of the wreck, and there waited. Presently he saw the boat coming like a phantom out of the gloom. It was quickly alongside, and the rescued people—twenty-five women and children—were transferred to the steamer, taken down to her cabin, and tenderly cared for. Making this transfer in such a sea was itself difficult in the extreme, and accompanied with great danger, but difficulty and danger were the rule that night, not the exception. All went well. The Aid, with the warrior-boat in tow, steamed back to windward of the wreck; then the lifeboat slipped the cable as before, and returned to the conflict, leaping over the seething billows to the field of battle like a warhorse refreshed.

The stirring scene was repeated with success. Forty women and children were rescued on the second trip, and put on board the steamer. Leaden daylight now began to dawn. Many hours had the "storm warriors" been engaged in the wild exhausting fight, nevertheless a third and a fourth time did they charge the foe, and each time with the same result. All the passengers were finally rescued and put on board the steamer.

But now arose a difficulty. The tide had been falling and leaving the wreck, so that the captain and crew determined to stick to her in the hope of getting her off, if the gale should abate before the tide rose again.

It was therefore agreed that the lifeboat should remain by her in case of accidents; so the exhausted men had to prepare for a weary wait in their wildly plunging boat, while the Aid went off with her rescued people to Ramsgate.

But the adventures of that night were not yet over. The tug had not been gone above an hour and a half, when, to the surprise of those in the lifeboat, she was seen returning, with her flag flying half mast high, a signal of recall to her boat. The lifeboat slipped from the side of the wreck and ran to meet her. The reason was soon explained. On his way back to Ramsgate the captain had discovered another large vessel on her beam-ends, a complete wreck, on that part of the sands named the Shingles. It was the Demerara, and her crew were still seen clinging to the quivering mast on which they had spent the livelong night.

More work for the well-nigh worn out heroes! Away they went to the rescue as though they had been a fresh crew. Dashing through the surf they drew near the doomed ship, which creaked and groaned when struck by the tremendous seas, and threatened to go to pieces every moment. The sixteen men on the mast were drenched by every sea. Several times that awful night they had, as it were, been mocked by false hopes of deliverance. They had seen the flashing of the rockets and faintly heard the thunder of the alarm-guns fired by the lightships. They had seen the lights of the steamer while she searched in vain for them on first reaching the sands, had observed the smaller light of the boat in tow, whose crew would have been so glad to save them, and had shouted in vain to them as they passed by on their errand of mercy to other parts of the sands, leaving them a prey to darkness and despair. But a merciful and loving God had seen and heard them all the time, and now sent them aid at the eleventh hour.

When the lifeboat at last made in towards them the ebb tide was running strongly, and, from the position of the wreck, it was impossible to anchor to windward and drop down to leeward in the usual fashion. They had, therefore, to adopt the dangerous plan of running with the wind, right in upon the fore-rigging, and risk being smashed by the mast, which was beating about with its living load like an eccentric battering-ram. But these Ramsgate men would stick at nothing. They rushed in and received many severe blows, besides dashing into the iron windlass of the wreck. Slowly, and one by one, the enfeebled men dropped from the mast into the boat. Sixteen—all saved! There was great shaking of hands, despite the tossings of the hungry surf, and many fervid expressions of thankfulness, as the sail was hoisted and the men of the Demerara were carried away to join the other rescued ones, who by that time thronged the little Aid almost to overflowing.

At Ramsgate that morning—the morning of the 4th—it was soon known to the loungers on the pier that the lifeboat was out, had been out all night, and might be expected back soon. Bright and clear, though cold, was the morn which succeeded that terrible night; and many hundreds of anxious, beating, hopeful hearts were on the lookout. At last the steamer and her warrior-boat appeared, and a feeling of great gladness seemed to spread through the crowd when it was observed that a flag was flying at the mast-head, a well-known sign of victory.

On they came, right gallantly over the still turbulent waves. As they passed the pier-heads, and the crowd of pale faces were seen gazing upwards in smiling acknowledgment of the hearty welcome, there burst forth a deep-toned thrilling cheer, which increased in enthusiasm as the extent of the victory was realised, and culminated when it became known that at one grand swoop the lifeboat, after a fight of sixteen hours, had rescued a hundred and twenty souls from the grasp of the raging sea!

Reader, there was many a heart-stirring incident enacted that night which I have not told you, and much more might be related of that great battle and glorious victory. But enough, surely, has been told to give you some idea of what our coast heroes dare and do in their efforts to rescue the perishing.

CHAPTER THREE.

LIGHT AND SHADE IN LIFEBOAT WORK.

But victory does not always crown the efforts of our lifeboats. Sometimes we have to tell of partial failure or defeat, and it is due to the lifeboat cause to show that our coast heroes are to the full as daring, self-sacrificing, and noble, in the time of disaster as they are in the day of victory.

A splendid instance of persevering effort in the face of absolutely insurmountable difficulty was afforded by the action of the Constance lifeboat, belonging to Tynemouth, on the night of the 24th November, 1864.

On that night the coast of Northumberland was visited by one of the severest gales that had been experienced for many years, and a tremendous sea was dashing and roaring among the rocks at the mouth of the Tyne. Many ships had sought refuge in the harbour during the day, but, as the shades of evening began to descend, the risk of attempting an entrance became very great. At last, as the night was closing in, the schooner Friendship ran on the rocks named the Black Middens.

Shortly afterwards a large steamer, the Stanley, of Aberdeen, with thirty passengers (most of whom were women), thirty of a crew, a cargo of merchandise, and a deckload of cattle, attempted to take the river. On approaching she sent up rockets for a pilot, but none dared venture out to her. The danger of putting out again to sea was too great. The captain therefore resolved to attempt the passage himself. He did so. Three heavy seas struck the steamer so severely as to divert her from her course, and she ran on the rocks close to the Friendship, so close that the cries of her crew could be heard above the whistling winds and thundering waves. As soon as she struck, the indescribable circumstances of a dread disaster began. The huge billows that had hitherto passed onward, heaving her upwards, now burst over her with inconceivable violence and crushed her down, sweeping the decks continuously—they rocked her fiercely to and fro; they ground her sides upon the cruel rocks; they lifted her on their powerful crests, let her fall bodily on the rocks, stove in her bottom, and, rushing into the hold, extinguished the engine fires. The sound of her rending planks and timbers was mingled with the piercing cries of the female passengers and the gruff shouting of the men, as they staggered to and fro, vainly attempting to do something, they knew not what, to avert their doom.

It was pitch dark by this time, yet not so dark but that the sharp eyes of earnest daring men on shore had noted the catastrophe. The men of the coastguard, under

Mr Lawrence Byrne, their chief officer, got out the rocket apparatus and succeeded in sending a line over the wreck. Unfortunately, however, owing to mismanagement of those on board the steamer, it proved ineffective. They had fastened the hawser of the apparatus to the forecastle instead of high up on the mast, so that the ropes became hopelessly entangled on the rocks. Before this entanglement occurred, however, two men had been hauled ashore to show the possibility of escape and to give the ladies courage. Then a lady ventured into the sling-lifebuoy, or cradle, with a sailor, but they stuck fast during the transit, and while being hauled back to the wreck, fell out and were drowned. A fireman then made the attempt. Again the cradle stuck, but the man was strong and went hand over hand along the hawser to the shore, where Mr Byrne rushed into the surf and caught hold of him. The rescuer nearly lost his life in the attempt. He was overtaken by a huge wave, and was on the point of being washed away when he caught hold of a gentleman who ran into the surf to save him.

The rocket apparatus having thus failed, owing to the simple mistake of those in the wreck having fastened the hawser *too low* on their vessel, the crew attempted to lower a boat with four seamen and four ladies in it. One of the davits gave way, the other swung round, and the boat was swamped. Three of the men were hauled back into the steamer, but the others perished. The men would not now launch the other boats. Indeed it would have been useless, for no ordinary boat could have lived in such a sea. Soon afterwards all the boats were washed away and destroyed, and the destruction of the steamer itself seemed about to take place every moment.

While this terrible fight for dear life was going on, the lifeboat-men were not idle. They ran out their good boat, the Constance, and launched her. And what a fearful launching that was! This boat belonged to the Institution, and her crew were justly proud of her.

According to the account given by her gallant coxswain, James Gilbert, they could see nothing whatever at the time of starting but the white flash of the seas as they passed over boat and crew, without intermission, twelve or thirteen times. Yet, as quickly as the boat was filled, she emptied herself through her discharging-tubes. Of these tubes I shall treat hereafter. Gilbert could not even see his own men, except the second coxswain, who, I presume, was close to him. Sometimes the boat was "driven to an angle of forty or forty-five degrees in clearing the rocks." When they were in a position to make for the steamer, the order was given to "back all oars and keep her end-on to the sea." The men obeyed; they seemed to be inspired with fresh vigour as they neared the wreck. Let Gilbert himself tell the rest of the story as follows.

"When abreast of the port bow, two men told us they had a rope ready on the starboard bow. We said we would be there in a moment. I then ordered the bow-man to be ready to receive the rope. As soon as we were ready we made two dashing strokes, and were under the bowsprit, expecting to receive the rope, when we heard a dreadful noise, and the next instant the sea fell over the bows of the Stanley, and buried the lifeboat. Every oar was broken at the gunwale of the boat, and the outer ends were swept away. The men made a grasp for the spare oars. Three were gone; two only remained. We were then left with the rudder and two oars. The next sea struck the boat almost over end on board the Friendship, the boat at the time being nearly perpendicular. We then had the misfortune to lose four of our crew. As the boat made a most fearful crash, and fell alongside the vessel, James Grant was, I believe, killed on the spot, betwixt the ship and the boat; Edmund Robson and James Blackburn were thrown out, Joseph Bell jumped as the boat fell. My own impression is that the men all jumped from the boat on to the vessel. We saw them no more. There were four men standing in a group before the mainmast of the schooner. We implored them to come into the boat, but no one answered."

Little wonder at that, James Gilbert! The massive wreck must have seemed — at least to men who did not know the qualities of a lifeboat — a surer foothold than the tossed cockleshell with "only two oars and a rudder," out of which four of her own gallant crew had just been lost. Even landsmen can perceive that it must have required much faith to trust a lifeboat in the circumstances.

"The next sea that struck the lifeboat," continues the coxswain, "landed her within six feet of the foundation-stone of Tynemouth Dock, with a quickness seldom witnessed. The crew plied the remaining two oars to leeward against the rudder and boathook. We never saw anything till coming near the three Shields lifeboats. We asked them for oars to proceed back to the Friendship, but they had none to spare."

Thus the brave Constance was baffled, and had to retire, severely wounded, from the fight. She drove, in her disabled and unmanageable condition, into the harbour. Of the four men thrown out of her, Grant and Robson, who had found temporary refuge in the wrecked schooner, perished. The other two, Bell and Blackburn, were buoyed up by their cork lifebelts, washed ashore, and saved. The schooner itself was afterwards destroyed, and her crew of four men and a boy were lost.

Meanwhile the screams of those on board of her and the Stanley were borne on the gale to the vast crowds who, despite darkness and tempest, lined the neighbouring cliffs, and the Shields lifeboats just referred to made gallant attempts to approach the wrecks, but failed. Indeed, it seemed to have been a rash attempt on the part of the noble fellows of the Constance to have made the venture at all.

The second cabin of the Stanley was on deck, and formed the bridge, or outlook. On this a number of the passengers and crew had taken refuge, but a tremendous sea carried it, and all its occupants, bodily away. After this the fury of the sea increased, and about an hour before midnight the steamer, with a hideous crash, broke in two amidships. The after part remained fast; the fore part swung round. All the people who remained on the after part were swept away and drowned. The new position into which the fore part of the wreck had been forced was so far an advantage to those who still clung to it, that the bows broke the first violence of the waves, and thus partially protected the exhausted people, thirty-five of whom still remained alive out of the sixty souls originally on board. Ten of these were passengers — two being ladies.

Meanwhile fresh preparations were being made by the rocket-men. Messengers had been sent in hot haste to Cullercoats for more rockets, those at Tynemouth having been exhausted. They arrived at five o'clock in the morning. By that time the tide had fallen considerably, admitting of a nearer approach to the wreck, and once more a gleam of hope cheered the hearts of the perishing as they beheld the fiery messenger of mercy rush fiercely towards them from the shore. But hope was still delayed. Four of the rockets missed. The fifth passed right over them, dropping the lifeline on the wreck, and drawing from the poor sufferers a feeble cheer, which was replied to lustily from the shore. This time, fortunately, no mistakes were made by those on board. The blocks and tackle were drawn out, the hawser on which the sling-lifebuoy traversed was fastened high up on the foremast to prevent the ropes fouling the rocks, as they had done on the first attempt; then the lifebuoy was run out, and, eventually, every soul was drawn in safety to the shore.

Thus did that battle end, with much of disaster and death to regret, indeed, but with upwards of thirty-five rescued lives to rejoice over.

I have now shown the action and bearing of our coast heroes, both in circumstances of triumphant victory and of partial success. Before proceeding to other matters it is well to add that, when intelligence of this disaster was telegraphed to the Lifeboat Institution, a new lifeboat was immediately forwarded to Tynemouth, temporarily to replace the damaged Constance. Instructions were given for the relief of the widows and children of the two lifeboat-men who had perished, and 26 pounds was sent to the crew of the boat. At their next meeting the committee of the Institution, besides recording their deep regret for the melancholy loss of life, voted 100 pounds in aid of a fund raised locally for the widows and seven children of the two men. They likewise bestowed their silver medal and a vote of thanks, inscribed on vellum, to Mr Lawrence Byrne, of the coastguard, in testimony of his gallant services on the occasion. Contributions were also raised by a local committee for the relief of the sufferers by these disasters, and a Volunteer Corps was formed to assist in working the rocket apparatus on future occasions of shipwreck.

Let me at this point earnestly request the reader who dwells in an *inland* home, and who never hears the roaring of the terrible sea, carefully to note that in this case it was *men of the coast* who did the work, and *people of the coast town* who gave subscriptions, who sympathised with sufferers, and raised a Volunteer Corps. Ponder this well, good reader, and ask yourself the question, "Is all as it should be here? Have I and my fellow-inlanders nothing to do but read, admire, and say, Well done?" A hint is sufficient at this point. I will return to the subject hereafter.

Sometimes our gallant lifeboat-men when called into action go through a very different and not very comfortable experience. They neither gain a glorious victory nor achieve a partial success, but, after all their efforts, risks, and exposure, find that their services are not required, and that they must return meekly home with nothing to reward them but an approving conscience!

One such incident I once had the opportunity of observing. I was living at the time—for purposes of investigation, and by special permission—on board of the Gull Lightship, which lies directly off Ramsgate Harbour, close to the Goodwin

Sands. It was in the month of March. During the greater part of my two weeks' sojourn in that lightship the weather was reasonably fine, but one evening it came on to blow hard, and became what Jack styles "dirty." I went to rest that night in a condition which may be described as semi-sea-sick. For some time I lay in my bunk moralising on the madness of those who choose the sea for a profession. Suddenly I was roused—and the seasickness instantly cured—by the watch on deck shouting down the hatchway to the mate, "South Sand Head Light is firing, sir, and sending up rockets!"

The mate sprang from his bunk—just opposite to mine—and was on the cabin floor before the sentence was well finished. Thrusting the poker with violence into the cabin fire, he rushed on deck. I jumped up and pulled on coat, nether garments, and shoes, as if my life depended on my speed, wondering the while at the poker incident. There was unusual need for clothing, for the night was bitterly cold.

On gaining the deck I found the two men on duty actively at work, one loading the lee gun, the other fitting a rocket to its stick. A few hurried questions by the mate elicited all that it was needful to know. The flash of a gun from the South Sand Head Lightship, about six miles distant, had been seen, followed by a rocket, indicating that a vessel had got upon the fatal sands in her vicinity. While the men were speaking I saw the flash of another gun, but heard no report, owing to the gale carrying the sound to leeward. A rocket followed, and at the same moment we observed the distress signal of the vessel in danger flaring on the southern tail of the sands, but very faintly; it was so far away, and the night so thick.

By this time our gun was charged and the rocket in position.

"Look alive, Jack; fetch the poker!" cried the mate, as he primed the gun.

I was enlightened as to the poker! Jack dived down the hatchway and next moment returned with that instrument red-hot. He applied it in quick succession to gun and rocket. A grand flash and crash from the first was followed by a blinding blaze and a whiz as the second sprang with a magnificent curve far away into surrounding darkness. This was our answer to the South Sand Head Lightship. It was, at the same time, our signal-call to the lookout on the pier of Ramsgate Harbour.

"That's a beauty!" said our mate, referring to the rocket. "Get up another, Jack. Sponge her well out, Jacobs; we'll give 'em another shot in a few minutes."

Loud and clear were both our signals, but four and a half miles of distance and a fresh gale neutralised their influence on that dark and dismal night. The lookout did not see them. In a few minutes the gun and rocket were fired again. Still no answering signal came from Ramsgate.

"Load the weather gun!" said the mate.

Jacobs obeyed, and I sought shelter under the lee of the weather bulwarks, for the wind seemed to be made of pen-knives and needles! The sturdy Gull straining and plunging wildly at her huge cables, trembled as our third gun thundered forth its summons, but the rocket struck the rigging and made a low, wavering flight. Another was therefore sent up, but it had scarcely cut its bright line across the sky when we observed the answering signal—a rocket from Ramsgate pier.

"That's all right now, sir; *our* work is done," said the mate to me, as he went below and quietly turned in, while the watch, having sponged out and re-covered the gun, resumed their active perambulations of the deck.

I confess that I felt somewhat disappointed at the sudden termination of the noise and excitement. I was told that the Ramsgate lifeboat could not well be out in less than an hour. There was nothing for it, therefore, but patience, so I turned in, "all standing," as sailors have it, with a request that I should be called when the lights of the tug should come in sight. Scarcely had I lain down, however, when the voice of the watch was heard shouting hastily, "Lifeboat close alongside, sir! Didn't see it till this moment. She carries no lights."

Out I bounced, minus hat, coat, and shoes, and scrambled on deck just in time to see a boat close under our stern, rendered spectrally visible by the light of our lantern. It was not the Ramsgate but the Broadstairs lifeboat, the men of which had observed our first rocket, had launched their boat at once, and had run down with the favouring gale.

"What are you firing for?" shouted the coxswain of the boat.

"Ship on the sands bearing south," replied Jack, at the full pitch of his stentorian voice.

The boat which was under sail, did not pause, and nothing more was said. With a magnificent rush it passed us, and shot away into the darkness. Our reply had been heard, and the lifeboat, steering by compass, went straight as an arrow to the rescue.

It was a thrilling experience to me! Spectral as a vision though it seemed, and brief almost as the lightning flash, its visit was the *real* thing at last. Many a time had I heard and read of our lifeboats, and had seen them reposing in their boat-houses, as well as out "for exercise," but now I had *seen* a lifeboat tearing before the gale through the tormented sea, sternly bent on the real work of saving human life.

Once again all became silent and unexciting on board the Gull, and I went shivering below with exalted notions of the courage, endurance, and businesslike vigour of our coast heroes. I now lay wakeful and expectant. Presently the shout came again.

"Tug's in sight, sir!"

And once more I went on deck with the mate.

The steamer was quickly alongside, heaving wildly in the sea, with the Ramsgate lifeboat "Bradford" in tow far astern. She merely slowed a little to admit of the same brief question and reply, the latter being repeated, as the boat passed, for the benefit of the coxswain. As she swept by us I looked down and observed that the ten men who formed her crew crouched flat on the thwarts. Only the steersman sat up. No wonder. It must be hard to sit up in a stiff gale with freezing spray, and sometimes heavy seas sweeping over one. I knew that the men were wide awake and listening, but, as far as vision went that boat was manned only by ten oilskin coats and sou'-westers!

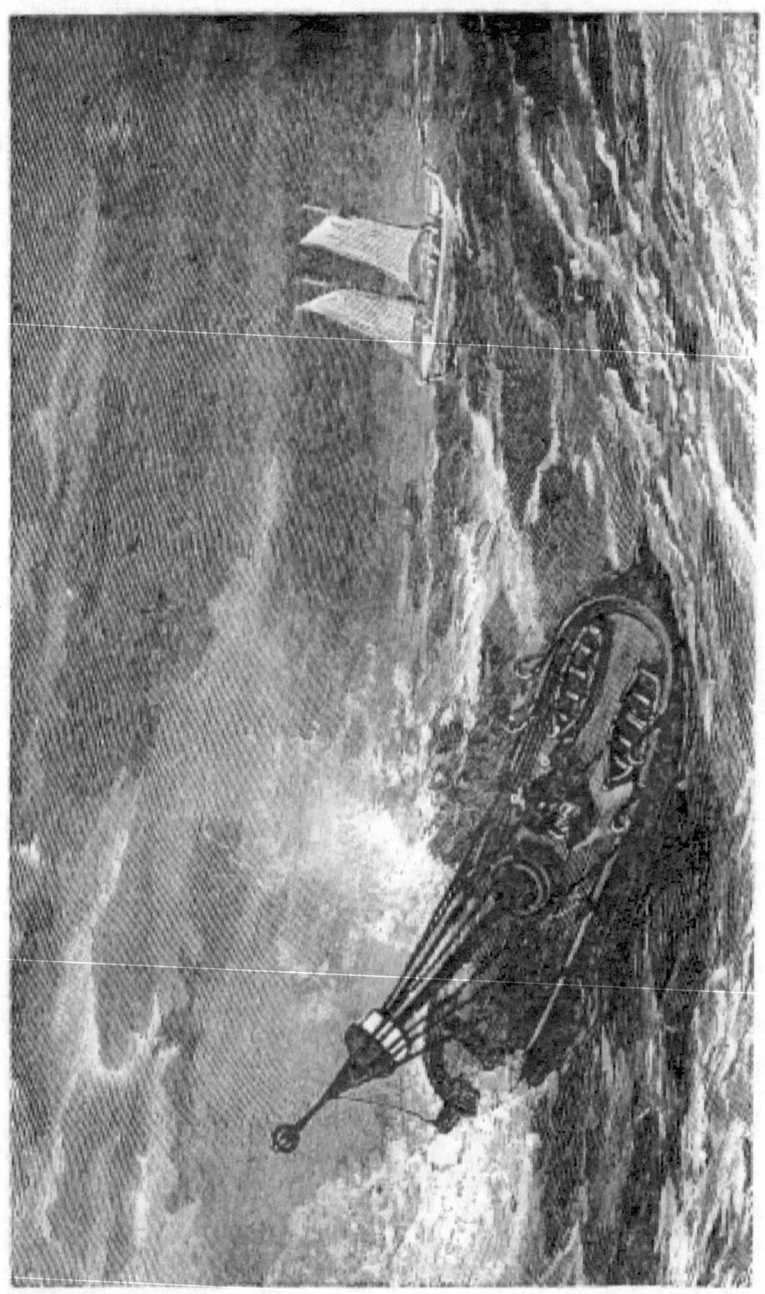

THE BROADSTAIRS LIFEBOAT PASSING THE "GULL" LIGHT.

A few seconds carried them out of sight, and thus, as regards the Gull Light-ship, the drama ended. There was no possibility of the dwellers in the floating lights hearing anything of the details of that night's work until the fortnightly visit of their "tender" should fall due, but next morning at low tide, far away in the distance, we could see the wreck, bottom up, high on the Goodwin Sands.

Afterwards I learned that the ship's crew had escaped in one of their own boats, and taken refuge in the South Sand Head Lightship, whence they were conveyed next day to land, so that the gallant men of Ramsgate and Broadstairs had all their toil and trouble for nothing!

Thus, you see, there are not only high lights and deep shadows, but also neutral tints in the various incidents which go to make up the grand picture of lifeboat work.

There is a Fund connected with the Broadstairs Lifeboat which deserves passing notice here. It was raised by the late Sir Charles Reed, in 1867, the proceeds to be distributed annually among the seamen who save life on that coast. The following particulars of this fund were supplied by Sir Charles Reed himself:—

"Eight boatmen of Broadstairs were interested in a lugger—the Dreadnought—which had for years done good service on the Goodwins. One night they went off in a tremendous sea to save a French barque; but though they secured the crew, a steam-tug claimed the prize and towed her into Ramsgate Harbour. The Broadstairs men instituted proceedings to secure the salvage, but they were beaten in a London law court, where they were overpowered by the advocacy of a powerful company. In the meantime they lost their lugger off the coast of Normandy, and in this emergency the lawyers they had employed demanded their costs. The poor men had no means, and not being able to pay they were taken from their homes and lodged in Maidstone Gaol. He (Sir Charles) was then staying in Broadstairs, and an appeal being made to him, he wrote to the 'Times', and in one week received nearly twice the amount required. The bill was paid, the men were liberated and brought home to their families, and the balance of the amount, a considerable sum, was invested, the interest to be applied to the rewarding of boatmen who, by personal bravery, had distinguished themselves by saving life on the coast."

CHAPTER FOUR.

CONSTRUCTION AND QUALITIES OF THE LIFE-BOAT.

In previous chapters enough has been told, I think, to prove that our lifeboats deserve earnest and thoughtful attention, not only as regards their work, but in reference to their details of construction. It has been said that the lifeboat possesses special qualities which distinguish it from all other boats. Chief among these are the self-righting and self-emptying principles. Stability, resulting from breadth of beam, etcetera, will do much to render a boat safe in rough seas and tempestuous weather, but when a boat has to face mighty rollers which turn it up until it stands straight on end, like a rearing horse, and even tumble it right over, or when it has to plunge into horrible maelstroms which seethe, leap, and fume in the mad contention of cross seas, no device that man has yet fallen upon will save it from turning keel up and throwing its contents into the water.

Instead therefore, of attempting to build a boat which cannot upset, men have deemed it wiser to attempt the construction of one which will not remain in that position, but which will, of necessity, right itself. The end aimed at has been achieved, and the boat now in use by the Lifeboat Institution is absolutely perfect in this respect. What more could be desired in any boat than that, after being upset, it should right itself in a *few seconds*, and empty itself of water in less than one minute?

A boat which does not right itself when overturned is only a lifeboat so long as it maintains its proper position on the water.

Let its self-emptying and buoyant qualities be ever so good, you have only to upset it to render it no better than any other boat;—indeed, in a sense, it is worse than other boats, because it leads men to face danger which they would not dare to encounter in an ordinary boat.

Doubtless, lifeboats on the non-self-righting principle possess great stability, and are seldom overturned; nevertheless they occasionally are, and with fatal results. Here is one example. In the month of January, 1865, the Liverpool lifeboat, when out on service, was upset, and seven men of her crew were drowned. This was not a self-righting boat, and it did not belong to the Lifeboat Institution, most of whose boats are now built on the self-righting principle. Moreover, the unfortunate men had not put on lifebelts. It may be added that the men who work the boats of the Institution are not allowed to go off without their cork lifebelts on. .

Take another case. On the 4th January, 1857, the Point of Ayr lifeboat, when under sail in a gale, upset at a distance from the land. The accident was seen from the shore, but no aid could be rendered, and the whole boat's crew—thirteen in number—were drowned. This boat was considered a good lifeboat, and doubtless it was so in many respects, but it was not a self-righting one. Two or three of the poor fellows were seen clinging to the keel for twenty minutes, by which time they became exhausted, were washed off, and, having no lifebelts on, perished.

Again in February, 1858, the Southwold lifeboat—a large sailing boat, esteemed one of the finest in the kingdom, but not on the self-righting principle—went out for exercise, and was running before a heavy surf with all sail set, when she suddenly ran on the top of a sea, turned broadside to the waves, and was upset. The crew in this case were fortunately near the shore, had on their lifebelts, and, although some of them could not swim, were all saved—no thanks, however, to their boat, which remained keel up—but three unfortunate gentlemen who had been permitted to go off in the boat without lifebelts, and one of whom was a good swimmer, lost their lives.

Let it be noted here that the above three instances of disaster occurred in the day time, and the contrast of the following case will appear all the stronger.

One very dark and stormy night in October, 1858, the small lifeboat of Dungeness put off through a heavy sea to a wreck three-quarters of a mile from the shore. Eight stout men of the coastguard composed her crew. She was a self-righting, self-emptying boat, belonging to the Lifeboat Institution. The wreck was reached soon after midnight, and found to have been abandoned. The boat, therefore, returned towards the shore. Now, there is a greater danger in rowing before a gale than in rowing against it. For the first half mile all went well, though the sea was heavy and broken, but, on crossing a deep channel between two shoals, the little lifeboat was caught up and struck by three heavy seas in succession. The coxswain lost command of the rudder, and she was carried away before a sea, broached to, and upset, throwing her crew out of her. *Immediately* she righted herself, cleared herself of water, and was brought up by her anchor which had fallen out when she was overturned. The crew meanwhile having on lifebelts, floated and swam to the boat, caught hold of the life-lines festooned round her sides, clambered into her, cut the cable, and returned to the shore in safety! What more need be said in favour of the self-righting boats?

The self-emptying principle is quite equal to the self-righting in importance.

In *every* case of putting off to a wreck in a gale, a lifeboat ships a great deal of water. In most cases she fills more than once. Frequently she is overwhelmed by tons of water by every sea. A boat full of water cannot advance, therefore baling becomes necessary; but baling, besides being very exhausting work, is so slow that it would be useless labour in most cases. Besides, when men have to bale they cannot give that undivided attention to the oars which is needful. To overcome this difficulty the self-emptying plan was devised.

As, I doubt not, the reader is now sufficiently interested to ask the questions, How are self-righting and self-emptying accomplished? I will try to throw some light on these subjects.

First, as to self-righting. You are aware, no doubt, that the buoyancy of our lifeboat is due chiefly to large air-cases at the ends, and all round the sides from stem to stern. The accompanying drawing and diagrams will aid us in the description. On the opposite page you have a portrait of, let us say, a thirty-three feet, ten-oared lifeboat, of the Royal National Lifeboat Institution, on its transporting carriage, ready for launching, and, on page 95, two diagrams representing respectively a section and a deck view of the same (Figures 1, 2, and 3).

The breadth of this boat is eight feet; its stowage-room sufficient for thirty passengers, besides its crew of twelve men—forty-two in all. It is double-banked; that is, each of its five banks, benches, or thwarts, accommodates two rowers sitting side by side. The lines festooned round the side dip into the water, so that anyone swimming alongside may easily grasp them, and in the middle part of the boat— just where the large wheels come in the engraving—two of the lines are longer than the others, so that a man might use them as stirrups, and thus be enabled to clamber into the boat even without assistance. The rudder descends considerably below the keel—to give it more power—and has to be raised when the boat is being launched.

The shaded parts of the diagrams show the position and form of the air-cases which prevent a lifeboat from sinking. The white oblong space in Figure 2 is the free space available for crew and passengers. In Figure 3 is seen the depth to which the air-chambers descend, and the height to which the bow and stern-chambers rise.

It is to these large air-chambers in bow and stern, coupled with great sheer— or rise fore and aft—of gunwale, and a very heavy keel, that the boat owes its self-righting power. The two air-chambers are rounded on the top. Now, it is obvious that if you were to take a model of such a boat, turn it upside down on a table, and try to make it rest on its two *rounded* air-chambers, you would encounter as much difficulty as did the friends of Columbus when they sought to make an egg stand on its end. The boat would infallibly fall to one side or the other. In the water the tendency is precisely the same, and that tendency is increased by the heavy iron keel, which drags the boat violently round to its right position.

The self-righting principle was discovered—at all events for the first time exhibited—at the end of last century, by the Reverend James Bremner, of Orkney. He first suggested in the year 1792 that an ordinary boat might be made self-righting by placing two watertight casks in the head and sternsheets of it, and fastening three hundredweight of iron to the keel. Afterwards he tried the experiment at Leith, and with such success that in 1810 the Society of Arts voted him a silver medal and twenty guineas. But nothing further was done until half a century later, when twenty out of twenty-four pilots lost their lives by the upsetting of the non-self-righting Shields lifeboat.

FIG. 1.–LIFEBOAT ON CARRIAGE.

Then (1850) the late Duke of Northumberland offered a prize of 100 guineas for the best lifeboat that could be produced. No fewer than 280 models and drawings were sent in, and the plans, specifications, and descriptions of these formed five folio manuscript volumes! The various models were in the shape of pontoons, catamarans or rafts, north-country cobles, and ordinary boats, slightly modified. The committee appointed to decide on their respective merits had a difficult task to perform. After six months' careful, patient investigation and experiment, they awarded the prize to Mr James Beeching, of Great Yarmouth. Beeching's boat, although the best, was not, however, deemed perfect.

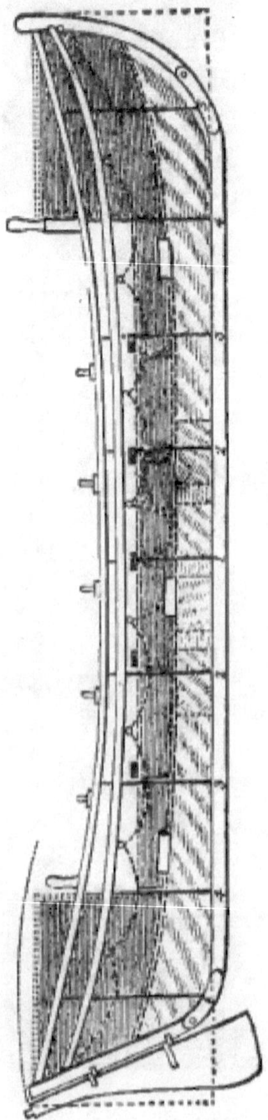

FIG. 2.–SECTION OF LIFEBOAT.

The committee therefore set Mr James Peake, one of their number, and assistant master-shipwright at Woolwich Dockyard, to incorporate as many as possible of the good qualities of all the other models with Beeching's boat. From time to time various important improvements have been made, and the result is the present magnificent boat of the Institution, by means of which hundreds of lives are saved every year.

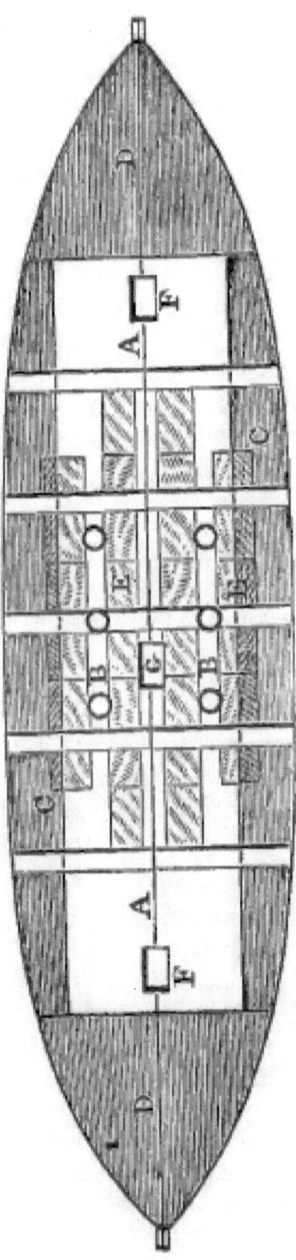

FIG. 3.–BIRD'S-EYE VIEW OF LIFEBOAT.

The self-discharge of water from a lifeboat is not so easy to explain. It will be the more readily comprehended if the reader understands, and will bear in remembrance, the physical fact that water will, and must, find its level. That is—no portion of water, small or great, in tub, pond, or sea, can for a moment remain above its flat and level surface, except when forced into motion, or commotion. Left to itself it infallibly flattens out, becomes calm, lies still in the lowest attainable position—in other words, finds its level. Bearing this in mind, let us look again at Figure 3.

The dotted double line about the middle of the boat, extending from stem to stern, represents the *floor* of the boat, on which the men's feet rest when standing or sitting in it. It also represents, or very nearly so, the waterline outside, that is, the depth to which the boat will sink when afloat, manned and loaded. Therefore, the *boat's floor* and the *ocean surface* are on the same level. Observe that! The space between the floor and the keel is filled up with cork or other ballast. Now, there are six large holes in the boat's floor—each hole six inches in diameter—into which are fitted six metal tubes, which pass down by the side of the cork ballast, and right through the bottom of the boat itself; thus making six large openings into the sea.

"But hallo!" you exclaim, "won't the water from below rush up through these holes and fill the boat?"

It will indeed rush up into these holes, but it will not fill the boat because it will have found its level—the level of ocean—on reaching the floor. Well, besides having reached its level, the water in the tubes has reached six valves, which will open downwards to let water out, but which won't open upwards to let it in. Now, suppose a huge billow topples into the boat and fills it quite full, is it not obvious that all the water in the boat stands *above* the ocean's level—being above the boat's floor? Like a wise element, it immediately seeks its own level by the only mode of egress—the discharging tubes; and when it has found its level, it has also found the floor of the boat. In other words, it is all gone! moreover, it rushes out so violently that a lifeboat, filled to overflowing, frees itself, as I have already said, in less than one minute!

The *buoyancy*, therefore, of a lifeboat is not affected for more than a few seconds by the tons of water which occasionally and frequently break into her. To prove this, let me refer you again to the account of the Constance, given by its gallant coxswain, as recorded in the third chapter. He speaks of the lifeboat being "buried," "sunk" by the wave that burst over the bow of the Stanley, and "immediately," he adds, "the men made a grasp for the spare oars!" There is no such remark as "when we recovered ourselves," etcetera. The sinking and leaping to the surface were evidently the work of a few seconds; and this is indeed the case, for when the force that sinks a lifeboat is removed, she rises that instant to the surface like a cork, and when she tumbles over she recovers herself with the agility of an acrobat!

The transporting-carriage is a most essential part of a lifeboat establishment, because wrecks frequently take place at some distance from a station, and prompt assistance is of the utmost importance in all cases of rescue. It is drawn by horses, and, with its exceedingly broad and strong wheels, can be dragged over any kind of road or across soft sand. It is always backed into the surf so deep that the boat may be launched from it, with her crew seated, and the oars out, ready to pull with might and main the instant the plunge is made. These first strokes of a lifeboat's crew are of immense importance. Want of union or energy on the part of steersman or crew at this critical point may be fatal. The boat must be made to cut the breakers end-on, so as to prevent her turning broadside on and being rolled back on the beach. Even after these initial strokes have been made successfully, there still remains the possibility of an unusually monstrous wave hurling the boat back end over end.

The boat resting on its carriage on the sands (Figure 1) shows the relative position of the two. It will be seen, from that position, that a very slight tip will suffice to cause the bow of the boat to drop towards the sea. As its keel rests on rollers, comparatively little force is required to launch it. Such force is applied by means of ropes attached to the stern, passing through pulleys at the outer end of the carriage, so that people on shore haul the ropes inland in order to force the boat off its carriage seaward.

Once the boat has got fairly over the surf and out upon the wild sea, her progress is comparatively safe, simple tugging against wind and sea being all that has to be done until the wreck is reached, where dangers of another kind await her.

I have now shown that the great qualities of our lifeboat are—*buoyancy*, or a tendency not to sink; *self-righting* power, or inability to remain upside down; *self-emptying* power, or a capacity to discharge any water that may get into it; and *stability*, or a tendency not to upset. The last quality I shall refer to, though by no means the least, is *strength*.

From what has been already written about lifeboats being hurled against wrecks and rocks, it must be evident that the strength of ordinary boats would not suffice.

In order to give them the requisite strength of frame for their tremendous warfare, they are built of the best Honduras mahogany, on what is known as the diagonal plan—that is, the boat has two distinct "skins" of planking, one set of planks being laid on in a diagonal position to the others. Moreover, these planks run from one gunwale round under the boat to the other gunwale, and have a complete layer of prepared canvas between them. Thus great strength and elasticity are combined, so that the boat can stand an inconceivable amount of battering on wreckage, rocks, or sand, without being destroyed.

That this is really so I will endeavour to prove by referring in the next chapter to a particular instance in which the great strength of one of our lifeboats was powerfully illustrated.

It may be added, in conclusion, that the oars of a lifeboat are short, and so made as to combine the greatest possible strength with lightness. They are fastened to the gunwale by short pieces of rope, and work in a moveable iron crutch on an iron thole-pin. Each boat is provided with a set of spare oars. Her equipment of compass, cables, grapnels, anchors, etcetera, is, as may be supposed, very complete, and she rides upon the storm in a rather gay dress of red, white, and blue, in order that she may be readily distinguished from other boats—her lower parts being white, her upper sides blue, and her line of "fender" all round being scarlet.

CHAPTER FIVE.
MORE TALES OF HEROISM.

If any one should doubt the fact that a lifeboat is *all but* indestructible, let that sceptical one read the following tale of wreck and rescue.

On a terrible night in the year 1857 a Portuguese brig struck on the Goodwin Sands, not far from the lightship that marks the northern extremity of those fatal shoals. A shot was fired, and a rocket sent up by the lightship. No second signal was needed. The Ramsgate men were, as usual, keeping a bright lookout. Instantly they jumped into the lifeboat, which lay calmly floating in the harbour alongside the pier. So eager were the men to engage in the deadly struggle that the boat was over-manned, and the last two who jumped in were obliged to go ashore again.

The tug *Aid* was all ready—according to custom—with steam up. She took the boat in tow and made for the mouth of the harbour. Staggering out in the teeth of tide and tempest they ploughed their way through a heavy cross sea, that swept again and again over them, until they reached the edge of the Goodwins. Here the steamer cast off the boat, and waited for her while she dashed into the surf, and bore the brunt of the battle alone.

It was a familiar proceeding to all concerned. Many a time before had the Ramsgate boat and steamer rescued men and women and little ones from the jaws of death on the Goodwins, but they were about to experience a few novelties that night.

It was very dark, so that the boat had much difficulty in finding the brig. On coming within about eighty yards of her they cast anchor and veered down under her lee. At first they were in hopes of getting the vessel off, and some hours were spent in vain attempts to do this, but the gale increased in fury; the brig began to break up. She rolled from side to side, and the yards swung wildly in the air. A blow from one of these yards would have stove the boat in, so the Portuguese crew—twelve men and a boy—were taken from the wreck, and the lifeboat-men endeavoured to push off.

All this time the boat had been floating in a basin worked in the sand by the motion of the wreck; but the tide had been falling, and when they tried to pull up to their anchor the boat struck heavily on the edge of this basin. They worked to get off the shoals with the energy of men who believe that their lives depend on their efforts. For a moment they succeeded in getting afloat, but again struck and remained fast.

Meanwhile the brig was lifted by each wave, that came rushing over the shoals like a mountain chain of snow, and let fall with a thundering crash. Her timbers began to snap like pipe-stems, and, as she worked nearer and nearer to the boat, the wildly-swaying yards threatened immediate destruction. The heavy seas flew continually over the lifeboat, so that passengers and crew could do nothing but hold on to the thwarts for their lives. At last the brig came so near that there was a stir among the men; they were preparing for the last struggle—some of them intending to leap into the rigging of the wreck and take their chance. But the coxswain shouted, "Stick to the boat, boys, stick to the boat!" and the men obeyed.

At that moment the boat lifted a little on the surf and grounded again. New hope was inspired by this. They pulled at the cable and shoved might and main with the oars. They succeeded in getting out of immediate danger, but still could not pull up to their anchor in the teeth of wind and tide. The coxswain then saw plainly that there was but one resource left—to cut the cable and drive away to leeward right across the Goodwin Sands, which at that place were two miles wide. But there was not yet sufficient water on the sands even for the attempting of that forlorn hope. As far as could be seen in that direction, ay, and far beyond the power of vision, there was nothing but a chaos of wild, tumultuous, whirling foam, without sufficient depth to float them over, so they held on, intending to wait till the tide, which had turned, should rise. Very soon, however, the anchor began to drag. This compelled them to hoist sail, cut the cable sooner than they had intended, and attempt to beat to windward—off the sands. It was in vain. A moment more, and they struck with tremendous force. A breaker came rolling towards them, filled the boat, caught her up like a plaything on its crest, and, hurling her a few yards onwards, let her fall with a shock that well-nigh tore every man out of her. Each successive breaker treated her in this way!

Those who dwell by the seashore know well those familiar ripples that mark the sands when the tide is out. On the Goodwins those ripples are gigantic banks, to be measured by feet, not by inches. I can speak from personal experience, having once visited the Goodwins and walked among the sand-banks at low water. From one to another of these banks this splendid boat was thrown. Each roaring surf caught it by the bow or stern, and, whirling it right round, sent it crashing on the next ledge. The Portuguese sailors gave up all hope and clung to the thwarts in silent despair, but the crew did not lose heart altogether. They knew the boat well, had often gone out to battle in her, and hoped that they might yet be saved, if they could only escape striking on the pieces of old wreck with which the sands were strewn.

Thus, literally, yard by yard, with a succession of shocks, that would have knocked any ordinary boat to pieces, did that lifeboat drive, during two hours, over two miles of the Goodwin Sands!

A thrilling and graphic account of this wreck and rescue is given in the Reverend John Gilmore's book, "Storm Warriors," in which he tells us that while this exciting work was going on, the *Aid* lay head to wind, steaming half power, and holding her own against the storm, waiting for her lifeboat, but no lifeboat returned to her, and her gallant captain became more and more anxious as time flew by. Could it be possible that her sturdy little comrade, with whom she had gone out to battle in hundreds of gales, was overcome at last and destroyed! They signalled again and again, but got no reply. Then, as their fears increased, they began to cruise about as near to the dangerous shoals as they dared—almost playing with death—as they eagerly sought for their consort. At last the conviction was forced upon them that the boat must have been stove by the wreck and swamped. In the midst of their gathering despair they caught sight of the lightship's bright beam, shining like a star of hope through the surrounding darkness. With a faint hope they made for the vessel and hailed her. "Have you seen anything of the lifeboat?" was the eager question. "Nothing! nothing!" was the sad reply. Back they went again to the place they had left, determined to cruise on, hoping against hope, till the night should pass away. Hour after hour they steamed hither and thither, with anxiously straining eyes. At last grey dawn appeared and the wreck became dimly visible. They made for it, and their worst fears were realised—the remnant of the brig's hull was there with ropes and wreckage tossing wildly round it—but no lifeboat!

Sadly they turned away and continued to search for some time in the faint hope that some of her crew might be floating about, buoyed up by their lifebelts, but none were found, and at last they reluctantly made for the harbour.

And when the harbour was gained what saw they there? The lifeboat! safe and sound, floating as calmly beside the pier as if nothing had happened! As the captain of the *Aid* himself said, he felt inclined at once to shout and cry for wonder, and we may be sure that his wonder was not decreased when he heard the lifeboat's story from the brave coxswain's lips—how that, after driving right across the sands, as I have described, they suddenly found themselves in deep water. That then, knowing the extremity of danger to be past, they had set the sails, and, soon after, had, through God's mercy, landed the rescued Portuguese crew in Ramsgate Harbour!

It must not be imagined, however, that such work as this can be done without great cost to those who undertake it.

Some of the men never recovered from the effects of that night's exposure. The gratitude of the Portuguese seamen was very great, as well as their amazement at such a rescue! It is recorded of them that, before arriving in the harbour, they were observed to be in consultation together, and that one who understood a little English spoke to one of the crew in an undertone.

"Coxswain," said the lifeboat man, "they want to give us all their money!"

"Yes, yes," cried the Portuguese interpreter, in broken English; "you have saved our lives! Thank you, thank you! but all we have is yours. It is not much, but you may take it between you." The amount was seventeen pounds!

As might have been expected, neither the coxswain nor his men would accept a penny of it.

This coxswain was Isaac Jarman, who for many years led the famous Ramsgate lifeboat into action, and helped to save hundreds of human lives. While staying at Ramsgate I had the pleasure of shaking the strong hard hand of Jarman, and heard some of his adventures from his own lips.

Now, from all that has been said, it will, I think, be seen and admitted that the lifeboats of the Institution are almost indestructible.

The *lifebelt*, to which reference has been so often made, deserves special notice at this point. The figure on the title-page shows its appearance and the manner in which it is worn. It was designed in 1854, by Admiral J.R. Ward, the Institution's chief inspector of lifeboats. Its chief quality is its great buoyancy, which is not only sufficient to support a man with head and shoulders above water when heavily clothed, but enables the wearer easily to support another person—the extra buoyancy being 25 pounds. Besides possessing several great advantages over other lifebelts, that of Admiral Ward is divided in the middle by a space, where the waistbelt is fastened. This permits of great freedom of action, and the whole machine is remarkably flexible. It is also very strong, forming a species of armour which protects the wearer from severe blows, and, moreover, helps to keep him warm.

It behoves me now to say a few words about the inventor of lifeboats. As has been told, our present splendid boat is a combination of all the good points and improvements made in such boats down to the present time. But the man who first thought of a lifeboat and invented one; who fought against apathy and opposition; who completed and launched his ark of mercy on the sea at Bamborough, in the shape of a little coble, in the year 1785, and who actually saved many lives therewith, was a London coachbuilder, *Lionel Lukin* by name.

Assuredly this man deserved the deepest gratitude of the nation, for his was the first lifeboat ever brought into action, and he inserted the small end of that wedge which we have been hammering home ever since, and which has resulted in the formation of one of the grandest, most thoroughly national and unsectarian of our charitable institutions.

Henry Greathead—a boatbuilder of South Shields—erroneously got the credit of this invention. Greathead was a noted improver and builder of lifeboats, and was well and deservedly rewarded for his work; but he was not the inventor. Lionel Lukin alone can claim that honour.

In regard to the men who man them, enough has been written to prove that they well deserve to be regarded as the heroes of the coast!

And let me observe in passing that there are also *heroines* of the coast, as the following extract from the Journal of the Institution will show. It appeared in the January number of 1865.

"Voted the Silver Medal of the Institution, and a copy of its vote of thanks on parchment, to Miss Alice R. Le Geyt, in admiration of her prompt and courageous conduct in rowing a small boat into the surf at the risk of her life, and rescuing two little boys who had fallen into the sea from the outer pier at Lyme Regis, Dorset, on the 4th August."

Again, in October, 1879, the Committee of the National Lifeboat Institution voted the Silver Medal of the Institution, and a copy of the vote inscribed on vellum, to Miss Ellen Francis Prideaux Brune, Miss Gertrude Rose Prideaux Brune, Miss Mary Katherine Prideaux Brune, Miss Beatrice May Prideaux Brune, and Miss Nora O'Shaughnessy, in acknowledgment of their intrepid and prompt services in proceeding through a heavy surf in their rowing-boat, and saving, at considerable risk of life, a sailor from a boat which had been capsized by a squall of wind off Bray Hill, Padstow Harbour, Cornwall, on the 9th August. When the accident occurred, the ladies' boat was being towed astern of a fishing-boat, and Miss Ellen Prideaux Brune, with great gallantry and determination, asked to be cast off, and, with her companions, she proceeded with all possible despatch to the rescue of the drowning sailor. All the ladies showed great courage, presence of mind, and marked ability in the management of their small boat. They ran great risk in getting the man into it, on account of the strong tide and sea on at the time.

So it would appear that the spirit of the far-famed Grace Darling has not yet departed from the land!

If heroism consists in boldly facing and successfully overcoming dangers of the most appalling nature, then I hold that thousands of our men of the coast—from Shetland to the Land's End—stand as high as do those among our soldiers and sailors who wear the Victoria Cross. Let us consider an example.

On that night in which the Royal Charter went down, there was a Maltese sailor on board named Joseph Rodgers, who volunteered to swim ashore with a rope. Those who have seen the effect of a raging sea even on a smooth beach, know that the power of the falling waves is terrible, and their retreating force so great that the most powerful swimmers occasionally perish in them. But the coast to which Rodgers volunteered to swim was an almost perpendicular cliff.

I write as an eye-witness, reader, for I saw the cliff myself, a few days after the wreck took place, when I went down to that dreary coast of Anglesea to identify the bodies of lost kindred. Ay, and at that time I also saw something of the awful aspect of loss by shipwreck. I went into the little church at Llanalgo, where upwards of thirty bodies lay upon the floor—still in their wet garments, just as they had been laid down by those who had brought them from the shore. As I entered that church one body lay directly in my path. It was that of a young sailor. Strange to say, his cheeks were still ruddy as though he had been alive, and his lips were tightly compressed—I could not help fancying—with the force of the last strong effort he had made to keep out the deadly sea. Just beyond him lay a woman, and beside her a little child, in their ordinary walking-dresses, as if they had lain down there and fallen asleep side by side. I had to step across these silent forms, as they lay, some in the full light of the windows, others in darkened corners of the little church, and to gaze earnestly into their dead faces for the lineaments of those whom I had gone to find—but I did not find them there. Their bodies were washed ashore some days afterwards. A few of those who lay on that floor were covered to hide the mutilation they had received when being driven on the cruel rocks. Altogether it was an awful sight—well fitted to draw forth the prayer, "God help and bless those daring men who are willing to risk their lives at any moment all the year round, to save men and women and little ones from such a fate as this!"

But, to return to Joseph Rodgers. The cliff to which he volunteered to swim was thundered on by seas raised by one of the fiercest gales that ever visited our shores. It was dark, too, and broken spars and pieces of wreck tossing about increased the danger; while the water was cold enough to chill the life-blood in the stoutest frame. No one knew better than Rodgers the extreme danger of the attempt, yet he plunged into the sea with a rope round his waist. Had his motive been self-preservation he could have gained the shore more easily without a rope; but his motive was not selfish — it was truly generous. He reached the land, hauled a cable ashore, made it fast to a rock, and began to rescue the crew, and I have no doubt that every soul in that vessel would have been saved if she had not suddenly split across and sunk. Four hundred and fifty-five lives were lost, but before the catastrophe took place *thirty-nine* lives were saved by the heroism of that Maltese sailor. The Lifeboat Institution awarded its gold medal, with its vote of thanks inscribed on vellum, and 5 pounds, to Rodgers, in acknowledgment of his noble conduct.

All round the kingdom the men are, as a rule, eager to man our lifeboats. Usually there is a *rush* to the work; and as the men get only ten shillings per man in the daytime, and twenty shillings at night, on each occasion of going off, it can scarcely be supposed that they do it only for the sake of the pay! True, those payments are increased on occasions of unusual risk or exposure; nevertheless, I believe that a worthier motive animates our men of the coast. I do not say, or think, that religious feeling is the cause of their heroism. With some, doubtless, it is; with others it probably is not; but I sincerely believe that the *Word of God* — permeating as it does our whole community, and influencing these men either directly or indirectly — is the cause of their self-sacrificing courage, as it is unquestionably the cause of our national prosperity.

CHAPTER SIX.

SUPPLIES A FEW POINTS FOR CONSIDERATION.

I have now somewhat to say about the Royal National Lifeboat Institution, which has the entire management and control of our fleet of 273 lifeboats. That Institution has had a glorious history. It was founded by Sir William Hillary, Baronet—a man who deserves a monument in Westminster Abbey, I think; for, besides originating the Lifeboat Institution, he saved, and assisted in saving, 305 lives, with his own hands!

Born in 1824, the Institution has been the means of saving no fewer than 29,608 lives up to the end of 1882.

At its birth the Archbishop of Canterbury presided; the great Wilberforce, Lord John Russell, and other magnates were present; the Dukes of Kent, Sussex, and other members of the Royal family, became vice-patrons, the Earl of Liverpool its president, and George the Fourth its patron. In 1850 good Prince Albert became its vice-patron, and her Majesty the Queen became, and still continues, a warm supporter and annual contributor. This is a splendid array of names and titles, but let me urge the reader never to forget that this noble Institution depends on the public for the adequate discharge of its grand work, for it is supported almost entirely by voluntary contributions.

The sole object of the Institution is to provide and maintain boats that shall save the lives of shipwrecked persons, and to reward those who save lives, whether by means of its own or other boats. The grandeur of its aim and singleness of its purpose are among its great recommendations.

When, however, life does not require to be saved, and when opportunity offers, it allows its boats to save property.

It saves—and rewards those who assist in saving—many hundreds of lives every year. Last year (1882) the number saved by lifeboats was 741, besides 143 lives saved by shore-boats and other means, for which rewards were given by the Institution; making a grand total of 884 lives saved in that one year. The number each year is often larger, seldom less. One year (1869) the rescued lives amounted to the grand number of 1231, and in the greater number of cases the rescues were effected in circumstances in which ordinary boats would have been utterly useless—worse than useless, for they would have drowned their crews. In respect of this matter the value of the lifeboat to the nation cannot be estimated—at least, not until we invent some sort of spiritual arithmetic whereby we may calculate the price of widows' and orphans' tears, and of broken hearts!

BATTLING THROUGH THE SURF.

But in regard to more material things it is possible to speak definitely.

It frequently happens in stormy weather that vessels show signals of distress, either because they are so badly strained as to be in a sinking condition, or so damaged that they are unmanageable, or the crews have become so exhausted as to be no longer capable of working for their own preservation. In all such cases the lifeboat puts off with the intention in the first instance of saving life. It reaches the vessel in distress; some of the boat's crew spring on board, and find, perhaps, that there is some hope of saving the ship. Knowing the locality well, they steer her clear of rocks and shoals. Being comparatively fresh and vigorous, they work the pumps with a will, manage to keep her afloat, and finally steer her into port, thus saving ship and cargo as well as crew.

Now let me impress on you that incidents of this sort are not of rare occurrence. There is no play of fancy in my statements; they happen every year. Last year (1882) twenty-three vessels were thus saved by lifeboat crews. Another year thirty-three, another year fifty-three, ships were thus saved. As surely and regularly as the year comes round, so surely and regularly are ships and property saved by lifeboats—saved *to the nation*! It cannot be too forcibly pointed out that a wrecked ship is not only an individual, but a national loss. Insurance protects the individual, but insurance cannot, in the nature of things, protect the nation. If you drop a thousand sovereigns in the street, that is a loss to you, but not to the nation; some lucky individual will find the money and circulate it. But if you drop it into the sea, it is lost not only to you, but to the nation, indeed to the world itself, for ever,—of course taking for granted that our amphibious divers don't fish it up again!

Well, let us gauge the value of our lifeboats in this light. If a lifeboat saves a ship worth ten or twenty thousand sovereigns from destruction, it presents that sum literally as a free gift to owners *and* nation. A free gift, I repeat, because lifeboats are provided solely to save life—not property. Saving the latter is, therefore, extraneous service. Of course it would be too much to expect our gallant boatmen to volunteer to work the lifeboats, in the worst of weather, at the imminent risk of their lives, unless they were also allowed an occasional chance of earning salvage. Accordingly, when they save a ship worth, say 20,000 pounds, they are entitled to put in a claim on the owners for 200 pounds salvage. This sum would be divided (after deducting all expenses, such as payments to helpers, hire of horses, etcetera) between the men and the boat. Thus—deduct, say, 20 pounds expenses leaves 180 pounds to divide into fifteen shares; the crew numbering thirteen men:—

13 shares to men at £12 each	£156
2 shares to boat	£24
Total	£180

Let us now consider the value of loaded ships.

Not very long ago a large Spanish ship was saved by one of our lifeboats. She had grounded on a bank off the south coast of Ireland. The captain and crew forsook her and escaped to land in their boats. One man, however, was inadvertently left on board. Soon after, the wind shifted; the ship slipped off the bank into deep water, and drifted to the northward. Her doom appeared to be fixed, but the crew of the Cahore lifeboat observed her, launched their boat, and, after a long pull against wind and sea, boarded the ship and found her with seven feet of water in the hold. The duty of the boat's crew was to save the Spanish sailor, but they did more, they worked the pumps and trimmed the sails and saved the ship as well, and handed her over to an agent for the owners. This vessel and cargo was valued at 20,000 pounds.

Now observe, in passing, that this Cahore lifeboat not only did much good, but received considerable and well-merited benefit, each man receiving 34 pounds from the grateful owners, who also presented 68 pounds to the Institution, in consideration of the risk of damage incurred to their boat. No doubt it may be objected that this, being a foreign ship, was not saved to *our* nation; but, as the proverb says, "It is not lost what a friend gets," and I think it is very satisfactory to reflect that we presented the handsome sum of 20,000 pounds to Spain as a free gift on that occasion.

This was a saved ship. Let us look now at a lost one. Some years ago a ship named the Golden Age was lost. It was well named though ill-fated, for the value of that ship and cargo was 200,000 pounds. The cost of a lifeboat with equipment and transporting carriage complete is about 650 pounds, and there are 273 lifeboats at present on the shores of the United Kingdom. Here is material for a calculation! If that single ship had been among the twenty-seven saved last year (and it *might* have been) the sum thus rescued from the sea would have been sufficient to pay for all the lifeboats in the kingdom, and leave 22,550 pounds in hand! But it was *not* among the saved. It was lost—a dead loss to Great Britain. So was the Ontario of Liverpool, wrecked in October, 1864, and valued at 100,000 pounds. Also the Assage, wrecked on the Irish coast, and valued at 200,000 pounds. Here are five hundred thousand pounds—half a million of money—lost by the wreck of these three ships alone. Of course, these three are selected as specimens of the most valuable vessels lost among the two thousand wrecks that take place *each year* on our coasts; they vary from a first-rate mail steamer to a coal coffin, but set them down at any figure you please, and it will still remain true that it would be worth our while to keep up our lifeboat fleet, for the mere chance of saving such valuable property.

But after all is said that can be said on this point, the subject sinks into insignificance when contrasted with the lifeboat's true work—the saving of human lives.

There is yet another and still higher sense in which the lifeboat is of immense value to the nation. I refer to the moral influence it exercises among us. If many hundreds of lives are annually saved by our lifeboat fleet, does it not follow, as a necessary consequence, that happiness and gratitude must affect thousands of hearts in a way that cannot fail to redound to the glory of God, as well as the good of man? Let facts answer this question.

We cannot of course, intrude on the privacy of human hearts and tell what goes on there, but there are a few outward symptoms that are generally accepted as pretty fair tests of spiritual condition. One of these is parting with money! Looking at the matter in this light, the records of the Institution show that thousands of men, women, and children, are beneficially influenced by the lifeboat cause.

The highest contributor to its funds in the land is our Queen; the lowliest a sailor's orphan child. Here are a few of the gifts to the Institution, culled almost at random from the Reports. One gentleman leaves it a legacy of 10,000 pounds.

Some time ago a sum of 5000 pounds was sent anonymously by "a friend." A hundred pounds comes in as a *second* donation from "a sailor's daughter." Fifty pounds come from a British admiral, and five shillings from "the savings of a child!" One-and-sixpence is sent by another child in postage-stamps, and 1 pound 5 shillings as the collection of a Sunday school in Manchester; 15 pounds from three fellow-servants; 10 pounds from a shipwrecked pilot, and 10 shillings, 6 pence from an "old salt." I myself had once the pleasure of receiving twopence for the lifeboat cause from an exceedingly poor but enthusiastic old woman! But my most interesting experience in this way was the receipt of a note written by a blind boy—well and legibly written, too—telling me that he had raised the sum of 100 pounds for the Lifeboat Institution.

And this beneficial influence of our lifeboat service travels far beyond our own shores. Here is evidence of that. Finland sends 50 pounds to our Institution to testify its appreciation of the good done by us to its sailors. President Lincoln, of the United States, when involved in all the anxieties of the great war between North and South, found time to send 100 pounds to the Institution in acknowledgment of services rendered to American ships in distress. Russia and Holland send naval men to inspect—not our armaments and *materiel* of hateful war, but—our lifeboat management! France, in generous emulation, starts a Lifeboat Institution of its own, and sends over to ask our society to supply it with boats—and, last, but not least, it has been said that foreigners, driven far out of their course and stranded, soon come to know that they have been wrecked on the British coast, by the persevering efforts that are made to save their lives!

And now, good reader, let me urge this subject on your earnest consideration. Surely every one should be ready to lend a hand to *rescue the perishing*! One would think it almost superfluous to say more. So it would be, if there were none who required the line of duty and privilege to be pointed out to them. But I fear that many, especially dwellers in the interior of our land, are not sufficiently alive to the claims that the lifeboat has upon them.

Let me illustrate this by a case or two—imaginary cases, I admit, but none the less illustrative on that account.

"Mother," says a little boy, with flashing eyes and curly flaxen hair; "I want to go to sea!"

He has been reading "Cook's Voyages" and "Robinson Crusoe," and looks wistfully out upon the small pond in front of his home, which is the biggest "bit of water" his eyes have ever seen, for he dwells among the cornfields and pastures of the interior of the land.

"Don't think of it, darling Willie. You might get wrecked,—perhaps drowned."

But "darling Willie" does think of it, and asserts that being wrecked is the very thing he wants, and that he's willing to take his chance of being drowned! And Willie goes on thinking of it, year after year, until he gains his point, and becomes the family's "sailor boy," and mayhap, for the first time in her life, Willie's mother casts more than a passing glance at newspaper records of lifeboat work. But she does no more. She has not yet been awakened. "The people of the coast naturally look after the things of the coast," has been her sentiment on the subject—if she has had any definite sentiments about it at all.

On returning from his first voyage Willie's ship is wrecked. On a horrible night, in the howling tempest, with his flaxen curls tossed about, his hands convulsively clutching the shrouds of the topmast, and the hissing billows leaping up as if they wished to lick him off his refuge on the cross-trees, Willie awakens to the dread reality about which he had dreamed when reading Cook and Crusoe. Next morning a lady with livid face, and eyes glaring at a newspaper, gasps, "Willie's ship—is—wrecked! five lost—thirteen saved by the lifeboat." One faint gleam of hope! "Willie may be among the thirteen!" Minutes, that seem hours, of agony ensue; then a telegram arrives, "*Saved, Mother—thank God,—by the lifeboat.*"

"Ay, thank God," echoes Willie's mother, with the profoundest emotion and sincerity she ever felt; but think you, reader, that she did no more? Did she pass languidly over the records of lifeboat work after *that* day? Did she leave the management and support of lifeboats to *the people of the coast*? I trow not. But what difference had the saving of Willie made in the lifeboat cause? Was hers the only Willie in the wide World? Are we to act on so selfish a principle, as that we shall decline to take an interest in an admittedly grand and good and national cause, until our eyes are forcibly opened by "our Willie" being in danger? Of course I address myself to people who have really kind and sympathetic hearts, but who, from one cause or another, have not yet had this subject earnestly submitted to their consideration. To those who have *no heart* to consider the woes and necessities of suffering humanity, I have nothing whatever to say,—except,—God help them!

Let me enforce this plea—that *inland* cities and towns and villages should support the Lifeboat Institution—with another imaginary case.

A tremendous gale is blowing from the south-east, sleet driving like needles— enough, almost, to put your eyes out. A "good ship," under close-reefed topsails, is bearing up for port after a prosperous voyage, but the air is so thick with drift that they cannot make out the guiding lights. She strikes and sticks fast on outlying sands, where the sea is roaring and leaping like a thousand fiends in the wintry blast. There are passengers on board from the Antipodes, with boxes and bags of gold-dust, the result of years of toil at the diggings. They do not realise the full significance of the catastrophe. No wonder—they are landsmen! The tide chances to be low at the time; as it rises, they awake to the dread reality. Billows burst over them like miniature Niagaras. The good ship which has for many weeks breasted the waves so gallantly, and seemed so solid and so strong, is treated like a cork, and becomes apparently an egg-shell!

Night comes—darkness increasing the awful aspect of the situation tenfold. What are boxes and bags of gold-dust now—now that wild despair has seized them all, excepting those who, through God's grace, have learned to "fear no evil?"

Suddenly, through darkness, spray, and hurly-burly thick, a ghostly boat is seen! The lifeboat! Well do the seamen know its form! A cheer arouses sinking hearts, and hope once more revives. The work of rescuing is vigorously, violently, almost fiercely begun. The merest child might see that the motto of the life-boat-men is "Victory or death." But it cannot be done as quickly as they desire; the rolling of the wreck, the mad plunging and sheering of the boat, prevent that.

A sturdy middle-aged man named Brown—a common name, frequently associated with common sense—is having a rope fastened round his waist by one of the lifeboat crew named Jones—also a common name, not seldom associated with uncommon courage. But Brown must wait a few minutes while his wife is being lowered into the boat.

"Oh! be careful. Do it gently, there's a good fellow," roars Brown, in terrible anxiety, as he sees her swung off.

"Never fear, sir; she's all right," says Jones, with a quiet reassuring smile, for Jones is a tough old hand, accustomed to such scenes.

Mrs Brown misses the boat, and dips into the raging sea.

"Gone!" gasps Brown, struggling to free himself from Jones and leap after her, but the grasp of Jones is too much for him.

"Hold on, sir? *she's* all right, sir, bless you; they'll have her on board in a minute."

"I've got bags, boxes, *bucketfuls* of gold in the hold," roars Brown. "Only save her, and it's all yours!"

The shrieking blast will not allow even *his* strong voice to reach the men in the lifeboat, but they need no such inducement to work.

"The gold won't be yours long," remarks Jones, with another smile. Neptune'll have it all to-night. See! they've got her into the boat all right, sir. Now don't struggle so; you'll get down to her in a minute. There's another lady to go before your turn comes.

During these few moments of forced inaction the self-possessed Jones remarks to Brown, in order to quiet him, that they'll be all saved in half an hour, and asks if he lives near that part of the coast.

"Live near it!" gasps Brown. "No! I live nowhere. Bin five years at the diggings. Made a fortune. Going to live with the old folk now—at Blunderton, far away from the sea; high up among the mountains."

"Hm!" grunts Jones. "Do they help to float the lifeboats at Blunderton?"

"The lifeboats? No, of course not; never think of lifeboats up there."

"Some of you think of 'em down *here*, though," remarks Jones. "Do *you* help the cause in any way, sir?"

"Me? No. Never gave a shilling to it."

"Well, never mind. It's your turn now, sir. Come along. We'll save you. Jump!" cries Jones.

And they do save him, and all on board of that ill-fated ship, with as much heartfelt satisfaction as if the rescued ones had each been a contributor of a thousand a year to the lifeboat cause.

"Don't forget us, sir, when you gits home," whispers Jones to Brown at parting.

And *does* Brown forget him? Nay, verily! He goes home to Blunderton, stirs up the people, hires the town-hall, gets the chief magistrate to take the chair, and forms a *Branch* of the Royal National Lifeboat Institution—the Blunderton Branch, which, ever afterwards, honourably bears its annual share in the expense, and in the privilege, of rescuing men, women, and little ones from the raging seas. Moreover, Brown becomes the enthusiastic secretary of the Branch. And here let me remark that no society of this nature can hope to succeed, unless its secretary be an enthusiast.

Now, reader, if you think I have made out a good case, let me entreat you to go, with Brown in your eye, "and do likewise."

And don't fancy that I am advising you to attempt the impossible. The supposed Blunderton case is founded on fact. During a lecturing tour one man—somewhat enthusiastic in the lifeboat cause—preached the propriety of inland towns starting Branches of the Lifeboat Institution. Upwards of half a dozen such towns responded to the exhortation, and, from that date, have continued to be annual contributors and sympathisers.

CHAPTER SEVEN.

THE LIFE-SAVING ROCKET.

We shall now turn from the lifeboat to our other great engine of war with which we do battle with the sea from year to year, namely, the Rocket Apparatus.

This engine, however, is in the hands of Government, and is managed by the coastguard. And it may be remarked here, in reference to coastguard men, that they render constant and effective aid in the saving of shipwrecked crews. At least one-third of the medals awarded by the Lifeboat Institution go to the men of the coastguard.

Every one has heard of Captain Manby's mortar. Its object is to effect communication between a stranded ship and the shore by means of a rope attached to a shot, which is fired over the former. The same end is now more easily attained by a rocket with a light rope, or line, attached to it.

Now the rocket apparatus is a little complicated, and ignorance in regard to the manner of using it has been the cause of some loss of life. Many people think that if a rope can only be conveyed from a stranded ship to the shore, the saving of the crew is comparatively a sure and easy matter. This is a mistake. If a rope—a stout cable—were fixed between a wreck and the shore, say at a distance of three or four hundred yards, it is obvious that only a few of the strongest men could clamber along it. Even these, if benumbed and exhausted—as is frequently the case in shipwreck—could not accomplish the feat. But let us suppose, still further, that the vessel rolls from side to side, dipping the rope in the sea and jerking it out again at each roll, what man could make the attempt with much hope of success, and what, in such circumstances, would become of women and children?

More than one rope must be fixed between ship and shore, if the work of saving life is to be done efficiently. Accordingly, in the rocket apparatus there are four distinct portions of tackle. First the *rocket-line*; second, the *whip*; third, the *hawser*; and, fourth, the *lifebuoy*—sometimes called the sling-lifebuoy, and sometimes the breeches-buoy.

The rocket-line is that which is first thrown over the wreck by the rocket. It is small and light, and of considerable length—the extreme distance to which a rocket may carry it in the teeth of a gale being between three and four hundred yards.

The whip is a thicker line, rove through a block or pulley, and having its two ends spliced together without a knot, in such a manner that the join does not check the running of the rope through the pulley. Thus the whip becomes a double line—a sort of continuous rope, or, as it is called, an "endless fall," by means of which the lifebuoy is passed to and fro between the wreck and shore.

The hawser is a thick rope, or cable, to which the lifebuoy is suspended when in action.

The lifebuoy is one of those circular lifebuoys—with which most of us are familiar—which hang at the sides of steamers and other vessels, to be ready in case of any one falling overboard. It has, however, the addition of a pair of huge canvas breeches attached to it, to prevent those who are being rescued from slipping through.

Let us suppose, now, that a wreck is on the shore at a part where the coast is rugged and steep, the beach very narrow, and the water so deep that it has been driven on the rocks not more than a couple of hundred yards from the cliffs. The beach is so rocky that no lifeboat would dare to approach, or, if she did venture, she would be speedily dashed to pieces—for a lifeboat is not *absolutely* invulnerable! The coastguardsmen are on the alert. They had followed the vessel with anxious looks for hours that day as she struggled right gallantly to weather the headland and make the harbour. When they saw her miss stays on the last tack and drift shoreward, they knew her doom was fixed; hurried off for the rocket-cart; ran it down to the narrow strip of pebbly beach below the cliffs, and now they are fixing up the shore part of the apparatus. The chief part of this consists of the rocket-stand and the box in which the line is coiled, in a peculiar and scarcely describable manner, that permits of its flying out with great freedom.

While thus engaged they hear the crashing of the vessel's timbers as the great waves hurl or grind her against the hungry rocks. They also hear the cries of agonised men and women rising even above the howling storm, and hasten their operations.

At last all is ready. The rocket, a large one made of iron, is placed in its stand, a *stick* and the *line* are attached to it, a careful aim is taken, and fire applied. Amid a blaze and burst of smoke the rocket leaps from its position, and rushes out to sea with a furious persistency that even the storm-fiend himself is powerless to arrest. But he can baffle it to some extent—sufficient allowance has not been made for the force and direction of the wind. The rocket flies, indeed, beyond the wreck, but drops into the sea, a little to the left of her.

"Another—look alive!" is the sharp order. Again the fiery messenger of mercy leaps forth, and this time with success. The line drops over the wreck and catches in the rigging. And at this point comes into play, sometimes, that ignorance to which I have referred—culpable ignorance, for surely every captain who sails upon the sea ought to have intimate acquaintance with the details of the life-saving apparatus of every nation. Yet, so it is, that some crews, after receiving the rocket-line, have not known what to do with it, and have even perished with the means of deliverance in their grasp. In one case several men of a crew tied themselves together with the end of the line and leaped into the sea! They were indeed hauled ashore, but I believe that most, if not all, of them were drowned.

Those whom we are now rescuing, however, are gifted, let us suppose, with a small share of common sense. Having got hold of the line, one of the crew, separated from the rest, signals the fact to the shore by waving a hat, handkerchief, or flag, if it be day. At night a light is shown over the ship's side for a short time, and then concealed. This being done, those on shore make the end of the line fast to the *whip* with its "tailed-block" and signal to haul off the line. When the whip is got on board, a *tally*, or piece of wood, is seen with white letters on a black ground painted on it. On one side the words are English—on the other French. One of the crew reads eagerly:—

"Make the tail of the block fast to the lower mast well up. If masts are gone, then to the best place you can find. Cast off the rocket-line; see that the rope in the block runs free, and show signal to the shore."

Most important cautions these, for if the tail-block be fastened too low on the wreck, the ropes will dip in the water, and perhaps foul the rocks. If the whip does not run free in the block it will jamb and the work will be stopped; and, if the signals are not attended to, the coastguardsmen may begin to act too soon, or, on the other hand, waste precious time.

But the signals are rightly given; the other points attended to, and the remainder of the work is done chiefly from the shore. The men there, attach the hawser to the whip, and by hauling one side thereof in, they run the other side and the hawser out. On receiving the hawser the crew discover another *tally* attached to it, and read:—

"Make this hawser fast about two feet above the tail-block. See all clear, and that the rope in the block runs free, and show signal to the shore."

The wrecked crew are quick as well as intelligent. Life depends on it! They fasten the end of the hawser, as directed, about two feet *above* the place where the tail-block is fixed to the stump of the mast. There is much shouting and gratuitous advice, no doubt, from the forward and the excited, but the captain and mate are cool. They attend to duty and pay no regard to any one.

Signal is again made to the shore, and the men of the coastguard at once set up a triangle with a pendent block, through which the shore-end of the hawser is rove, and attached to a double-block tackle. Previously, however, a block called a "traveller" has been run on to the hawser. This block travels on and *above* the hawser, and from it is suspended the lifebuoy. To the "traveller" block the whip is attached; then the order is given to the men to haul, and away goes the lifebuoy to the wreck, run out by the *men on shore.*

When it arrives at the wreck the order is, "Women first." But the women are too terrified, it may be, to venture. Can you wonder? If you saw the boiling surf the heaving water, the roaring and rushing waves, with black and jagged rocks showing here and there, over which, and partly through which, they are to be dragged, you would respect their fears. They shrink back: they even resist. So the captain orders a 'prentice boy to jump in and set them the example. He is a fine, handsome boy, with curly brown hair and bright black eyes. He, too, hesitates for a moment, but from a far different motive. If left to himself he would emulate the captain in being that proverbial "last man to quit the wreck," but a peremptory order is given, and, with a blush, he jumps into the bag, or breeches, of the buoy, through which his legs project in a somewhat ridiculous manner. A signal is then made to the shore. The coastguardsmen haul on the whip, and off goes our 'prentice boy like a seagull. His flight is pretty rapid, considering all things. When about half-way to land he is seen dimly in the mist of spray that bursts wildly around and over him. Those on the wreck strain their eyes and watch with palpitating hearts. The ship has been rolling a little. Just then it gives a heavy lurch shoreward, the rope slackens, and down goes our 'prentice boy into the raging sea, which seems to roar louder as if in triumph! It is but for a moment, however. The double-block tackle, already mentioned as being attached to the shore-end of the hawser, is manned by strong active fellows, whose duty it is to ease off the rope when the wreck rolls seaward, and haul it in when she rolls shoreward, thus keeping it always pretty taut without the risk of snapping it.

A moment more and the 'prentice is seen to emerge from the surf like a true son of Neptune; he is seen also, like a true son of Britain, to wave one hand above his head, and faintly, through driving surf and howling gale, comes a cheer. It is still more faintly replied to by those on the wreck, for in his progress the boy is hidden for a few seconds by the leaping spray; but in a few seconds more he is seen struggling among the breakers on the beach. Several strong men are seen to join hands and advance to meet him. Another moment, and he is safe on shore, and a fervent "Thank God!" bursts from the wrecked crew, who seem to forget themselves for a moment as they observe the waving handkerchiefs and hats which tell that a hearty cheer has greeted the rescued sailor boy.

There is little tendency now to hesitation on the part of the women, and what remains is put to flight by certain ominous groans and creakings, that tell of the approaching dissolution of the ship.

One after another they are lifted tenderly into the lifebuoy, and drawn to land in safety, amid the congratulations and thanksgivings of many of those who have assembled to witness their deliverance. It is truly terrible work, this dragging of tender women through surf and thundering waves; but it is a matter of life or death, and even the most delicate of human beings become regardless of small matters in such circumstances.

But the crew have yet to be saved, and there are still two women on board—one of them with a baby! The mother—a thin, delicate woman—positively refuses to go without her babe. The captain knows full well that, if he lets her take it, the child will be torn from her grasp to a certainty; he therefore adopts a seemingly harsh, but really merciful, course. He assists her into the buoy, takes a quick turn of a rope round her to keep her in, snatches the child from her arms, and gives the signal to haul away. With a terrible cry the mother holds out her arms as she is dragged from the bulwarks, then struggles to leap out, but in vain. Another wild shriek, with the arms tossed upwards, and she falls back as if in a fit.

"Poor thing!" mutters the captain, as he gazes pitifully at the retreating figure; "but you'll soon be happy again. Come, Dick, get ready to go wi' the child next trip."

Dick Shales is a huge hairy seaman, with the frame of an elephant, the skin of a walrus, and the tender heart of a woman! He glances uneasily round.

"There's another lady yet, sir."

"You obey orders," says the captain, sternly.

"I never disobeyed orders yet, sir, and I won't do it now," says Dick, taking the baby into his strong arms and buttoning it up tenderly in his capacious bosom.

As he speaks, the lifebuoy arrives again with a jovial sort of swing, as if it had been actually warmed into life by its glorious work, and had come out of its own accord.

"Now, then, lads; hold on steady!" says Dick, getting in, "for fear you hurt the babby. This is the first time that Dick Shales has appeared on any stage wotsomediver in the character of a woman!"

Dick smiles in a deprecating manner at his little joke as they haul him off the wreck. But Dick is wrong, and his mates feel this as they cheer him, for many a time before that had he appeared in woman's character when woman's work had to be done.

The captain was right when he muttered that the mother would be "soon happy again." When Dick placed the baby—wet, indeed, but well—in its mother's arms, she knew a kind of joy to which she had been a stranger before—akin to that joy which must have swelled the grateful heart of the widow of Nain when she received her son back from the dead.

The rest of the work is soon completed. After the last woman is drawn ashore the crew are quickly rescued—the captain, of course, like every true captain, last of all. Thus the battle is waged and won, and nothing is left but a shattered wreck for wind and waves to do their worst upon.

The rescued ones are hurried off to the nearest inn, where sympathetic Christian hearts and hands minister to their necessities. These are directed by the local agent for that admirable institution, the Shipwrecked Fishermen's and Mariners' Society—a society which cannot be too highly commended, and which, it is well to add, is supported by voluntary subscriptions.

Meanwhile the gallant men of the coastguard, rejoicing in the feeling that they have done their duty so well and so successfully, though wet and weary from long exposure and exertion, pack the rocket apparatus into its cart, run it back to its place of shelter, to be there made ready for the next call to action, and then saunter home, perchance to tell their wives and little ones the story of the wreck and rescue, before lying down to take much-needed and well-earned repose.

Let me say in conclusion that hundreds of lives are saved in this manner *every* year. It is well that the reader should bear in remembrance what I stated at the outset, that the Great War is unceasing. Year by year it is waged. There is no prolonged period of rest. There is no time when we should forget this great work; but there are times when we should call it specially to remembrance, and bear it upon our hearts before Him whom the wind and sea obey.

When the wild storms of winter and spring are howling; when the frost is keen and the gales are laden with snowdrift; when the nights are dark and long, and the days are short and grey—then it is that our prayers should ascend and our hands be opened, for then it is that hundreds of human beings are in deadly peril on our shores, and then it is that our gallant lifeboat and rocket-men are risking life and limb while fighting their furious Battles with the Sea.

THE END.

Lector House believes that a society develops through a two-fold approach of continuous learning and adaptation, which is derived from the study of classic literary works spread across the historic timeline of literature records. Therefore, we aim at reviving, repairing and redeveloping all those inaccessible or damaged but historically as well as culturally important literature across subjects so that the future generations may have an opportunity to study and learn from past works to embark upon a journey of creating a better future.

This book is a result of an effort made by Lector House towards making a contribution to the preservation and repair of original ancient works which might hold historical significance to the approach of continuous learning across subjects.

HAPPY READING & LEARNING!

LECTOR HOUSE LLP
E-MAIL: lectorpublishing@gmail.com